Praise for *The Biochar Handbook*

"A brief yet thorough introduction to the history, science, and practicalities of biochar-making and its tremendous potential to restore degraded soils. Highly recommended."

—**DAVID R. MONTGOMERY,** author of *Dirt* and *Growing a Revolution*

"This book is an absolute tour de force. It answers so many commonly asked questions about biochar in easy-to-understand prose. Everyone from newbies to those who have been in the biochar industry for years will learn a lot from reading this book. Recipes for bokashi, compost, toothpaste, and more showcase how biochar can be used both inside and outside the home."

—**KATHLEEN DRAPER,** coauthor of *Burn*; former International Biochar Initiative Board Chair

"Climate repair is the greatest challenge we have ever confronted, and it falls upon our generation, right now, no excuses. A formidable scholar and mechanical engineer, Kelpie Wilson has a way with words that few with either of those skills can equal. I would read her just to enjoy her style and humor, even if the subject were not so urgent and the stakes so astronomical."

—**ALBERT BATES,** 2024 EcoHero of the Year; author of *The Paris Agreement* and *The Biochar Solution*; coauthor of *Burn*

"Finally! A book on biochar that is not merely evangelizing, but written by someone with experience, who has read and done the research in order to learn the actual facts about what biochar is, what it does, and how it can be used at any scale of agriculture. It's good to see this subject finally addressed in a practical, no-nonsense way. Thumbs up!"

—**MARK SHEPARD,** author of *Restoration Agriculture* and *Water for Any Farm*

"A balanced, intelligent, and practical guide to one of the most promising ventures in the modern appropriate technology scene, Kelpie Wilson's book shows its author's detailed knowledge and extensive experience with biochar on every page. I would not be at all surprised to see *The Biochar Handbook* turn into the classic text in this burgeoning field."

—**JOHN MICHAEL GREER,**
author of *Green Wizardry* and *The Druid Path*

"This is by far the best book on biochar I've ever seen! We've all been needing this book—it's scientifically rigorous, well-written, and highly visual. If you don't have it already, get yourself a copy today!"

—**MATT POWERS,**
author of *The Regenerative Soil Trilogy*

The Biochar Handbook

The Biochar Handbook

A PRACTICAL GUIDE TO MAKING AND USING BIOACTIVATED CHARCOAL

Kelpie Wilson

Chelsea Green Publishing
White River Junction, Vermont
London, UK

Project Manager: Natalie Wallace
Developmental Editor: Ben Trollinger
Copy Editor: Will Solomon
Proofreader: Angela Boyle
Indexer: WordCo Indexing Services, Inc.
Designer: Melissa Jacobson
Page Layout: Abrah Griggs

Printed in the United States of America.
First printing June 2024.
10 9 8 7 6 5 4 3 2 1   24 25 26 27 28

Our Commitment to Green Publishing
Chelsea Green sees publishing as a tool for cultural change and ecological stewardship. We strive to align our book manufacturing practices with our editorial mission and to reduce the impact of our business enterprise in the environment. We print our books using vegetable-based inks whenever possible. This book may cost slightly more because it was printed on paper from responsibly managed forests, and we hope you'll agree that it's worth it. *The Biochar Handbook* was printed on paper supplied by Versa that is certified by the Forest Stewardship Council.®

Library of Congress Cataloging-in-Publication Data
Names: Wilson, Kelpie, 1956- author.
Title: The biochar handbook : a practical guide to making and using bioactivated charcoal / Kelpie Wilson
Description: First edition | White River Junction, Vermont : Chelsea Green Publishing, 2024 | Includes bibliographical references and index
Identifiers: LCCN 2024008461 (print) | LCCN 2024008462 (ebook) | ISBN 9781645022305 (paperback) | ISBN 9781645022312 (ebook)
Subjects: LCSH: Biochar. | Soil amendments. | Ashes as fertilizer.
Classification: LCC TP248.B55 W57 2024 (print) | LCC TP248.B55 (ebook) | DDC 631.8/6—dc23/eng/20240314
LC record available at https://lccn.loc.gov/2024008461
LC ebook record available at https://lccn.loc.gov/2024008462

Chelsea Green Publishing
White River Junction, Vermont USA
London, UK
www.chelseagreen.com

FSC
www.fsc.org
MIX
Paper from responsible sources
FSC® C005010

Contents

Introduction

"All that is very well," answered Candide, "but let us cultivate our garden."

—Voltaire, *Candide*

Dependence begets subservience and venality, suffocates the germ of virtue, and prepares fit tools for the designs of ambition.

—Thomas Jefferson

Biochar—what is it? After more than fifteen years of full-time work with biochar, my quick answer is that biochar is just a funny name for charcoal that you can add to soil—that also happens to sequester carbon from the atmosphere. Of course, biochar is much more than that, but the bottom line for me, and why I have stuck with it for so long, is that biochar has been a beautiful way for me to cultivate my own garden. By that, I don't mean just the garden that grows my food, but also the garden of my relationships in my community of place, enhanced by the network of biochar enthusiasts around the world whom I have come to know.

I have been a biochar consultant since 2012, helping companies and organizations understand biochar applications and markets. Before that I worked for the International Biochar Initiative for four years as an editor and project developer, helping the organization get off the ground. Now I teach workshops and online courses about making and using biochar, and I manufacture and sell the Ring of Fire biochar kiln.

I first heard of biochar in 2006 when it was known only as *terra preta*, the black earth of the Amazon. I was writing a regular column on energy, climate, and the environment for an online news site and I had been mostly focused on the peak oil story. Oil industry analysts were bringing to light the true state of the oil industry and the speed of oil depletion. According to some industry insiders, the world had reached the peak of conventional oil production globally on Thanksgiving Day, 2005. Going forward, oil would be produced with an ever-decreasing ERoEI (energy returned on energy invested), and following a long, bumpy plateau of price rises and demand destruction, oil and gas production would decline.

During my time as a journalist, I learned more about the dependence of our food system on cheap fossil fuels. In his book *Eating Fossil Fuels*, Dale Allen Pfeiffer shows how utterly dependent modern agriculture is on fossil fuels, not just for the machinery that plants and harvests but for the energy to irrigate fields and, most importantly, for fertilizers. About 30 percent of farm energy goes to fertilizer, much of which is made from natural gas. Like oil, natural gas is becoming increasingly expensive. Without gasoline, we might not drive cars, but without fertilizer, we might not eat.

When I heard about the mysterious terra preta soils of the Amazon that were seemingly able to grow crops without inputs, I felt relief. Perhaps we wouldn't have to starve when we ran out of oil. As it turns out, making and using biochar in my garden has been a huge boost to my own food security, and I can now grow an increasing proportion of what I consume using only biochar compost and small amounts of organic fertilizers.

I got the message early in life that industrial civilization could well have a limited tenure on the planet. There were gas lines in America in the 1970s when I was learning to drive. What good is being able to drive if there's no gas to put in the car? Somehow, I thought it was up to me to do something about it. It was certainly hard for me to have faith that government and society would tackle the problem of energy depletion, since it was rarely discussed or even acknowledged on the news or in my classes at school.

In my attempts to do something about the problem of industrial civilization, I have tried many different things. I became an auto mechanic and started growing gardens. I became a mechanical engineer and worked on alternative engines for power generation. I became a tree hugger and got involved in politics to change public policy and stop clear-cutting of ancient forests. I became a writer and wrote about peak oil, food security, sustainable forestry, biological diversity, climate change, and energy for news outlets and magazines.

Self-reliance is a value that becomes essential in bad times, but even in good times, it pays to be able to do for yourself as much as you can. Biochar has been a big help in my personal quest to feed myself and my family, take care of our waste, and manage the land around us. Biochar has also helped me stay healthy. The purpose of this book is to share with you what I have learned. Most of the tips and techniques for making and using biochar are things I have tried myself, or that I have personally observed. I will also discuss some techniques that I want to try, that I think will help me and my community.

I live in Oregon, where increasing temperatures, drought, and a long history of fire suppression have engendered a new era of megafires. My home is surrounded by forests, and I have seen the effects firsthand. Soon after I learned about terra preta, the US Forest Service did a thinning project on land next to my property. The workers cut small trees and low-hanging branches that could become ladder fuels into the forest canopy. They built hundreds of small burn piles scattered over forty acres and came back in the fall and lit them on fire on a rainy day. When the piles were out, I walked out into the forest to have a look and found that many of them had a layer of charcoal left along with the ash. The rain had saved some of the charcoal from burning up to ash.

I wondered if I could use this to make my own terra preta, so I shoveled up a few buckets of it and started using it in my garden. Ever since, I have enthusiastically explored different ways of making biochar in the woods using material that needs to be thinned to reduce the wildfire threat. In fact, it may have been that thinning project next to my house that saved it from the Slater Fire of 2020. The fire came right up to my driveway but did not burn the house.

I was working with a crew of volunteers from the Long Tom Watershed Council on a thinning project near Eugene, Oregon, in 2018 when we got the news about the Camp Fire that had ignited early that morning in the dry hills above Paradise, California. High winds blasted the flames through the town, incinerating most of the houses and killing more than eighty people. This was a wake-up call like no other. We needed to redouble our efforts to get more of the vital thinning work done. I hoped that making biochar from burn piles could help motivate people to do it.

I soon had an opportunity to help support people in Paradise with my biochar kilns. Stephen Feher, a professor at Butte Community College, had been teaching his students about biochar and starting biochar projects at local orchards. Stephen and his wife Elizabeth lost their house in the fire and were helping with

Figure 0.1. Not every tree burns to ash in a forest fire. Paradise homeowners were left with a lot of work to clean up the debris and process dead trees into firewood and biochar.

Figure 0.2. Stephen and Elizabeth Feher in their revived garden, now enriched with biochar. Their fruit trees were the only living things on their property that survived the fire.

the community recovery effort. Fire in a forest does not burn every tree to ash, and Paradise was now full of dead trees and brush that needed to be removed. The truck traffic to haul all that material away on the one main road in and out of the community would be disruptive, so most of it would have to be treated in place. Stephen invited me to teach a series of workshops on how to use my small kilns to cleanly process the debris into biochar. Soils had suffered from being burned, losing a lot of their organic matter. Biochar could help with soil recovery as well.

Forest fires are challenging for us, but for nature they are business as usual. Fire destroys but it also renews, and people in Paradise were enthusiastic about making biochar and seeing something good come out of the devastation. Our workshops brought thirty to forty people out for four days in a row to help their neighbors clean up properties and learn about biochar. Stephen and Elizabeth have now rebuilt their house and even revived their garden, adding biochar to the soil.

Of course, making biochar in the woods or in a backyard is not the only way to do it. There are many ways to make and use biochar, at different scales. A large biomass energy plant like the Biomass One power plant in Medford, Oregon, can make 3,500 to 4,000 tons of biochar a year while producing 30 megawatts of electricity. The wood for this power plant comes mostly from local sawmills, city green waste, and logging slash from nearby timber sales. Transporting forestry debris any distance longer than about 50 miles is not economical. This leaves a lot of stranded biomass that will never make it to an industrial facility, and provides an opportunity for making biochar in place on these remote sites.

Different regions have different patterns of biomass distribution on the landscape, whether from forestry, agriculture, industry, or landscaping. Biomass is always going to be widely distributed and difficult to gather up and transport. Every time you touch a piece of wood or a corn cob or a nut shell to gather it and move it somewhere else, it costs money. Biochar and bioenergy production facilities must also be distributed across the landscape, and specifically designed to handle the types of feedstocks that are locally available. Biomass is not a uniform commodity like coal or oil, and it does not scale up the same way. Instead, biochar and bioenergy facilities need to scale *out* across the landscape in a wide distribution. But this has not happened yet and the biochar industry struggles to find funding for thousands of small-scale facilities that are appropriately sized and designed for local and regional conditions, as all of our institutions are still hung up on the idea that bigger is better and that the economics only make sense when you can scale it up in a centralized facility.

As the biochar industry seeks capital to build plants, it also looks for markets for the product. This is difficult because not many people understand the value of biochar yet. Partly this is because biochar is most valuable when used at the start of a cascade of values that build on each other, and the cascade starts with a waste product. For instance, we can use biochar in animal barns where it improves animal health and controls odors. Biochar-amended manure makes a superior, high-nutrient compost. Applied to soil, a farmer will see benefits of lower water use, less fertilizer needed, better plant disease resistance, and higher nutrient value in the food. However, the farmer might not see huge improvements in yield, as compared to the current practice of chemical agriculture. In order for biochar to show a clear value proposition, we have to count every single value in the cascade, from improved waste management with less pollution to better quality food. This gets complicated and requires a lot of education of markets and a different perspective on what constitutes real value. These are problems for regenerative agriculture in general.

Biochar has been a challenging sell to farmers who have very tight profit margins and other constraints related to their dependence on big corporations for seed, fertilizer, and chemicals. For this reason, biochar proponents have researched other markets for biochar besides agriculture, including carbon removal credits for fixing carbon dioxide into stable carbon and keeping it out of the atmosphere. It turns out that the stable carbon found in biochar has many other uses in waste management, pollution control, water filtration, health and beauty, and construction materials. Besides carbon, biochar fixes:

- Drought conditions by holding water in soil
- Fertilizer shortages by holding nutrients in soil
- Dead soil by improving conditions for beneficial soil microbes
- Smelly manure and compost by supporting good microbes, balancing the carbon-to-nitrogen ratio, and retaining nutrients
- Polluted soil by immobilizing heavy metals and other contaminants
- Flooding by improving rainwater infiltration
- Eutrophication by absorbing nitrogen in water bodies
- Drinking water by filtering out contaminants
- Building material impacts when used as a substitute for resource-intensive components of asphalt, concrete, and other building materials

- Human health and wellbeing when used as an ingredient in health and beauty products and for management of human sanitation
- Animal health when used as an animal feed supplement and for manure management and sanitation
- Forest fires by converting excess fuel loads to water-holding soil

I cover most of these topics in this book to one degree or another, but the depth of information on each topic is limited by my own knowledge and experience. This biochar handbook is just one piece of what should eventually become an encyclopedia of ideas and methods for making and using biochar, with contributions from thousands of people who are using and innovating biochar methods around the world. In fact, major pieces of it already exist in documented practices both new and old, and can be found online and in research articles. I encourage you to explore the references in this book to learn more about the topics that interest you.

Ultimately, there is room for many more biochar handbooks to compile this knowledge and bring it forward. For instance, the use of biochar in cement, asphalt, and other building materials is not covered at all in this book. Albert Bates and Kathleen Draper's book *Burn: Using Fire to Cool the Earth* offers an excellent introduction to the subject, but eventually, as knowledge and experience accumulate, we will need a comprehensive handbook on how to use biochar in building materials.

Having a deeper and more comprehensive knowledge base about biochar that is widely distributed to the public will help build biochar markets. But here is the real beauty of biochar: We don't need to wait for a biochar industry to fully develop, although we will celebrate when it does. Anyone with a backyard can start making their own biochar and experimenting with ways to use it.

For me, it's got to be about what works in my community and my garden. I have to start from the ground where I grow my food. With that being the case, this book is focused on my experience here in the backwoods of Oregon, working with friends and neighbors and trying to be more self-reliant on an acre and a half of poor, rocky soil on a terrace above the Illinois River, surrounded by a temperate rainforest that is suffering from drought and mismanagement. I will also try to put my experience in the context of historic and current global and regional problems that can be ameliorated with biochar.

This biochar handbook will be useful for anyone who gardens, farms, raises animals, manages forests, or manages waste, at any scale—from a household, to a watershed, to a city. It will also be helpful for those who are not directly involved in any of these activities, but who want to understand the promise and potential of biochar to fix our relationship to the four elements of fire, water, earth, and sky, as we continue our journey here on planet Earth.

CHAPTER ONE

Biochar in History

Like many human discoveries, biochar has likely been invented, lost, and re-invented multiple times. You may have heard about the Amazonian terra preta soils, but did you know that such "dark earths," as they are known, are also found in Africa, Asia, and Europe? Even more astounding, to me at least, was that biochar was a big topic of discussion in the early industrial economies of Europe and America and that it almost revolutionized the way cities treated their sewage. No less a person than Justus Liebig, the father of agricultural chemistry, was a major player in the debate.

Analysis of "dark earths" found around the world shows that biochar was often mixed with human, animal, and kitchen wastes.[1] Later reports, including agricultural texts and journals from seventeenth-century Japan and nineteenth-century Europe and America, include numerous testimonials to the deodorizing power of charcoal and the high fertilizer value of manures composted with charcoal. As sanitation systems developed in the growing cities of the nineteenth century, advocates proposed that charcoal would convert human waste to valuable fertilizer in a sanitary and odorless fashion, while preventing the pollution of rivers and the loss of valuable nutrients. These advocates looked to the centuries-old systems in China and Japan for recycling urban nutrients for return to agricultural fields.

Farmers of Forty Centuries

Biochar was introduced to the modern world largely as a result of the discovery of the human-created (anthropogenic) soils in the Amazon, known as terra preta. The genesis of these soils began at least 2,000 to 2,500 years ago, if not earlier.[2] While geologists had been aware of these unusually black soils since the nineteenth century, it was not until the 1960s that a team led by the Dutch soil scientist Wim Sombroek began investigating them.[3] Today, researchers have concluded that terra preta soils were created through the process of waste disposal. Formed by the combination of charcoal, ash, food scraps, and human excreta, the expansion of these fertile soils facilitated the expansion of the human population in the Amazon.[4] Before the arrival of Europeans and their diseases (which decimated the Indigenous people), the settlements along certain parts of the Amazon River and its tributaries were concentrated on high river bluffs above the floodplain. This is where the terra preta soils are found.

Some anthropologists estimate that there were as many as ten million people living in the region before Columbus.[5] One can speculate that as the population expanded and the amount of human excreta and food scraps grew, it would have become ever more important to sanitize and deodorize waste. It makes sense that people would use charcoal and ash from cooking fires to control odor in food scraps and human excreta. As these were deposited near the settlements, the soil grew richer and more fertile. It's also possible that the improved soil in turn enabled more crops to be grown and more people could be supported. This is just one example of the sort of "virtuous cycle" (as opposed to "vicious cycle") that biochar can instigate.

Recent investigations into tropical regions in other parts of the world have found "anthropogenic dark earths" similar to the Amazonian terra preta. Researchers identified several sites in Borneo that shared many characteristics with terra preta soils, including high carbon content, high fertility, and a concentration along riversides.[6] An ongoing survey of dark earths in Africa has so far identified terra preta analogues at 134 sites in Liberia.[7] In Liberia, Guinea, Ghana, and Sierra Leone, anthropogenic dark earths are described as "commonplace." Dark earths are being formed in these regions today through the ongoing addition to soil of various forms of char and ash along with crop waste, cooking waste, and human excreta.

There is also a set of traditional practices that involves burning biomass in pits and trenches covered with dirt—directly in the fields—in order to improve soil fertility. These practices mix charcoal and ash into the soil. The process of heating the soil also affects nutrient cycling and physicochemical properties of the mineral soil by changing the clay structures. The use of covered biomass-burning trenches has been reported by aid workers in the Batibo region of Cameroon.[8] Biomass charring directly in farm fields is also still practiced in India, Bhutan, and until recently in Spain.[9]

Many traditional farming practices use fire to clear land ("slash and burn") or remove crop stubble. These practices are potentially a source of soil charcoal. However, in most cases the amount of charcoal is small compared to the amount of ash. Also, this sort of biomass burning does not involve the use of charcoal and ash to treat wastes such as food scraps, manure, or excreta.

The traditional farming practices of China, Japan, and Korea recycled massive amounts of human waste, ash, crop residue, and other biomass into agricultural fields. In 1909, the American agriculturalist F. H. King embarked on an eight-month tour of China, Japan, and Korea in order to view and document agricultural practices. The resulting book, *Farmers of Forty Centuries*, has become an agricultural classic.[10] Part of King's purpose in the book was to contrast the enduring agriculture of Asia with what he viewed as destructive and wasteful practices then advocated by the US Department of Agriculture.[11] King declared that one "of the most remarkable agricultural practices adopted by any civilized people is the centuries-long and well-nigh universal conservation and use of all human waste in China, Korea and Japan, turning it to marvelous account in the maintenance of soil fertility and in the production of food."[12] As an indicator of the commercial value of this human waste, he found that the city of Shanghai sold concessions to waste haulers, charging one contractor $31,000 in gold for the right to collect 78,000 tons of human waste for sale to farmers outside the city.[13]

King found compost making to be a high art in Japan, where prizes were offered in each county for the best compost. Winners at the county level went on to compete for a prize for best compost in the prefecture.[14] Although he did not specifically describe the use of charcoal in these composts, he observed that ash materials were added in large amounts. Most likely, ash included a proportion of cinders or charcoal. Moved by the thrift and care for conservation of nutrients that he observed on his travels, King expressed his frustration with the wasteful practices of his own country:

> When we reflect upon the depleted fertility of our own older farm lands, comparatively few of which have seen a century's service, and upon the enormous quantity of mineral fertilizers which are being applied annually to them in order to secure paying yields, it becomes evident that the time is here when profound consideration should be given to the practices the Mongolian race has maintained through many centuries.[15]

Contrasting these Asian practices with those in the United States, he said, "The rivers of North America are estimated to carry to the sea more than 500 tons of phosphorus with each cubic mile of water. To such loss modern civilization is adding that of hydraulic sewage disposal. . . ."[16]

Some historical uses of charcoal in Japanese agriculture are documented in a paper by Ogawa and Okimori titled "Pioneering Works in Biochar Research, Japan."[17] They describe a Japanese agricultural encyclopedia published in 1697 that gives instructions for making biochar compost: "After charring all waste, concentrated excretions should be mixed with it and stocked for a while. When you apply this manure to the fields, it is efficient for yielding any crop." Ogawa and Okimori claim that biochar has been in used in Asia since ancient times, and that rice husk charcoal has been used since the beginning of rice cultivation. Wood charcoal was not generally used in agriculture as it was too valuable as fuel. The article also reports:

> The practice of using rice husk charcoal mixed with excreta had been very popular in wheat cultivation until about 100 years ago. There were double benefits, which are that the charcoal can absorb and retain chemical nutrients as well as deodorize excreta. However, this method was so commonplace for local people that scientists rarely paid enough attention to investigating it.

Beginning in the 1980s, Ogawa and other scientists began to research these traditional uses of charcoal in agriculture in order to learn more about their effects. This revival of agricultural charcoal in Japan has been an important stimulus for the current attention to biochar.

One might expect that China would also possess historical texts on agricultural charcoal, but none have yet been published or described in English. Perhaps, as in Japan, the techniques were too common and widespread to attract much attention. However, archeological investigations may tell the tale. In 2011, the International Biochar Initiative sent me to China to address a meeting of biochar researchers about our development of biochar standards. One of the scientists at the meeting told me that thick layers of charcoal had been discovered in an ancient rice paddy, prompting him to speculate that these were analogous to terra preta.

Europe also has examples of ancient and historic use of charcoal in fields. Large extents of plaggen soils characterized by char particles and evidence of dung and other wastes can be found in Holland and elsewhere in northern Europe. Wim Sombroek, the Dutch soil scientist who is most responsible for initiating the scientific investigation of terra preta, recognized the similarity of the terra preta to the plaggen soils of his homeland:

> Wim Sombroek learned about soil as a child, during the hongerwinter—the Dutch wartime famine of 1944–45, in which 20,000 or more people died. His family survived on the harvest from a minute plot of plaggen soil: land enriched by generations of careful fertilization. If his ancestors had not taken care of their land, he once told me, the whole family might have died.[18]

While the scientific investigations into terra preta are very recent, there exists an extensive body of literature on the use of charcoal for both agriculture and sanitation beginning in the nineteenth century in Europe and America. It all began with Justus Liebig, the scientist we organic growers blame today for the dominance of chemical agriculture. As a scientist though, Liebig followed the evidence and was willing to change his mind when presented with new data. By the end of his life, Liebig was back on the right track.

Justus Liebig and the Overthrow of the Humus Theory

The science of agricultural charcoal in the West began with the German chemist Justus von Liebig. Liebig questioned the prevailing theory that humus, or black soil, was essential for plant growth because it imparted a vital life force to plants.

The idea that black soil was fertile soil had ancient roots. The Black Virgin images found in churches throughout Europe may represent the vital principle of the fertile black soil, an object worthy of veneration (the word "veneration" comes from the goddess Venus), since it was the source of bread and life for humanity.[19] Believers in humus thought that the black soil contained an organic life force or "vitalism" that could not be derived from dead, inorganic chemicals. This theory was based on the well-known fact that "virgin" soil from recently cleared forests was black and fertile.

Figure 1.1. Black Virgin at the Cathedral of Santiago de Compostela, Galicia, Spain. Photo by Cro Magnon / Alamy

Liebig reasoned that this must be wrong. Humus was clearly the product of plant decay, so how could the first plants on Earth have ever become established if humus was necessary? Supporting his theory, Liebig found that plants could be grown in pure charcoal—a "dead" chemical matter with no vital force. Vitalists countered that plants needed humus as a source of carbon, but Liebig believed that humus was too insoluble to provide

Figure 1.2. Plants can grow in pure biochar with just a bit of chemical NPK fertilizer.

the needed carbon. After all, charcoal was even more recalcitrant than humus and plants could grow in it. Therefore, he concluded that a plant that grows in charcoal must get its carbon from carbon dioxide in the air. Our Umpqua Biochar Education Team (UBET) replicated this experiment. We found that plants will grow in pure biochar with just a bit of chemical fertilizer and no other carbon source.

His scientific colleagues soon conceded that Liebig had "overthrown" the humus theory in favor of the chemical theory of agriculture. Through Liebig's experimental work, he also established the famous "law of the minimum," which states that plant growth is constrained by the least-available nutrient in the soil. These discoveries spurred a growing fertilizer industry that mined and shipped huge amounts of guano, bonemeal, lime, and other fertilizers from all parts of the world to fertilize the fields of Europe and eliminate the need for crop rotations and fallow periods to replenish the soil. Eventually, chemical processes were developed to manufacture fertilizers such as superphosphate and ammonia. Liebig's chemical theory of agriculture was as key to the Industrial Revolution as Watt's steam engine. Liebig certainly meant it to be. In his book *Agricultural Chemistry*, he wrote, "Perfect agriculture is the true foundation of all trade and industry."

But the success of the chemical fertilizer industry was not universally approved. Liebig himself had great misgivings about importation of fertilizers and the consequences for food security, "in the event of war with America when supplies of guano would cease."[20]

Imports were especially troublesome given the growing issue of town sewage disposal. Increasingly, London and other cities were flushing waste to the sea that had formerly been used to fertilize crops. Liebig's voice was one of many that decried the wastage of "night soil" (a term for human excreta). Liebig had read accounts of Chinese agriculture that led him to declare: "But how infinitely inferior is the agriculture of Europe to that of China! The Chinese are the most admirable gardeners and trainers of plants . . . the agriculture of their country is the most perfect in the world." Perfect because the Chinese understood the importance of the "most important of all manures," human excrement. Liebig said, "Indeed so much value is attached to the influence of human excrements by these people, that laws of the state forbid that any of them should be thrown away, and reservoirs are placed in every house, in which they are collected with the greatest care."[21]

Liebig would periodically try to influence policymakers who governed the development of the sewage and waste treatment systems in the great cities of Europe but ultimately to little effect.

The Passion for Charcoal

Liebig was a devoted experimentalist and he soon realized that humus did have a very important function in soil: It could absorb and hold chemical fertilizers that would otherwise leach away before plant roots could take them up. He compared the absorptivity of humus to that of charcoal and its ability, "surpassing all others," to absorb ammonia.

Liebig cited a series of hothouse experiments that a colleague had conducted using mixtures of charcoal in potted plants that gave fantastic results:

> An addition of charcoal, for example, to vegetable mould, appeared to answer excellently for the Gesneria and Gloxinia and also for the tropical Aroideae with tuberous roots. The two first soon excited the attention of connoisseurs, by the great beauty of all their parts and their general appearance.

The hothouse experiments also revealed that charcoal could have dramatic effects on the health of plants:

> Pure charcoal acts excellently as a means of curing unhealthy plants. A Dorianthes excelsa, for example, which had been drooping for three years, was rendered completely healthy in a very short time by this means. An orange tree, which had the very common disease in which the leaves become yellow, acquired within four weeks its healthy green color, when the upper surface of the earth was removed from the pot in which it was contained, and a ring of charcoal of an inch in thickness strewed in its place around the periphery of the pot. The same was the case with the Gardenia.[22]

Liebig did not have much more to say about these hothouse results with charcoal—he cited them to prove his point about the superfluity of humus, but popular journals for horticulturalists and farmers soon took notice of these experiments and published descriptions, articles, and testimonials on the beneficial effects of charcoal in soil. Many of the writers also observed that crops grown on old charcoal manufacturing sites, or on the kitchen middens of Indian villages, were especially abundant.

An anonymous farmer writing in an 1856 issue of *The New Jersey Farmer* reported:

> For two years past I have used some fifty loads each season of refuse charcoal, and being fully convinced that it pays, I wish to recommend it to my brother farmers. I have tried it on grass, corn and potatoes—have tried it alone and in the compost heap, and in all situations it has proved faithful to its trust. As a top dressing for grass, it gives a green color and luxuriant growth. Applied to half an acre of early potatoes the last summer, the yield was 75 bushels of as fine healthy potatoes as could be desired, that sold readily for one dollar per bushel, and yielded the best profit of anything raised on the farm.[23]

Another farmer writing in the journal of the New York State Agricultural Society described the use of charcoal for holding water in soil:

> In the midst of the disastrous drouth of last summer, while crossing a field in Moriah . . . I observed a lot with its surface deeply and singularly blackened. Upon inspection I found it thickly strewn with pulverized charcoal. The field presented a rich verdure, strongly contrasting with the parched and blighted aspect of the adjacent country.[24]

The entry on agricultural charcoal in the 1855 *Cyclopedia of Agriculture* runs to ten pages and covers many uses. On odor it says,

> Charcoal also possesses the property of absorbing and retaining the odoriferous and coloring principles of most organic substances. . . . From this deodorizing property, charcoal is frequently mixed with night soil, and other decaying manures; which it keeps free from smell, and at the same time aids in preserving, by absorbing the gases which would otherwise escape.[25]

Liebig's Warnings

It is interesting to note that while industry quickly found the profits to be made in mining and manufacturing chemical fertilizers, it never grasped the need for replenishing soil carbon, either in the form of humus or charcoal. Despite the interest expressed in the journals and amongst knowledgeable farmers and entrepreneurs, a biochar industry never took hold. The reasons were certainly economic in the main. More than one enthusiast echoed the complaint made in the 1847 volume of the *American Journal of Agriculture and Science*: "The use of charcoal as a fertilizer is generally well known. Its expense, however, often precludes its use."[26]

Perhaps if Liebig or another scientist of his stature had advocated the true importance of soil carbon, a biochar industry might have developed. But Liebig never did discover that soils could become depleted of carbon. He was convinced that soil carbon would always be replenished by the action of plants taking carbon from the atmosphere: "A certain quantity of carbon is taken every year from the forest or meadow, in the form of wood or hay, and, in spite of this, the quantity of carbon in the soil augments; it becomes richer in humus."[27]

Liebig did not live to see the widespread use of the ammonia fertilizers that have stimulated the rapid oxidation of soil carbon across the globe. On the other hand, Liebig did observe that adding manure to promote plant growth could eventually deplete soils of other chemical nutrients. He cited the example of a vineyard at Bingen on the Rhine that was heavily manured, greatly increasing plant growth. But the effect did not last because the soil was soon depleted of potash, removed with the harvest and not returned to the soil by the particular manure used. Yields soon plummeted and were worse than before the fertilization. Here was the law of the minimum at work, with a twist. For the first time it was apparent that fertilizer could actually harm soil.[28]

Liebig was very astute in another observation that was connected to some larger criticisms he had of the European system of agriculture. In contrast with China and Japan, where every bit of nutrient was returned to fields as human waste, Europe fertilized its fields with cattle droppings and dumped the waste from cities in rivers or other unproductive places. The problem was not just the loss of nutrients from sewage. Liebig realized that the cattle droppings came from pastures that would eventually become depleted because the deep roots of pasture grasses took minerals from the subsoil, and the minerals were transferred to the fields to grow grain that was exported as food to cities. While this practice might work for a long time, eventually the subsoil would become exhausted of minerals and the animals fed by pasture grasses would no longer produce manure sufficient to fertilize the grain.[29]

A *Poudrette* of No Account

Liebig felt that the best long-term solution to soil fertility was to adopt some form of the Chinese system. He said: "When human excrements are treated in a proper manner, so as to remove the moisture which they contain without permitting the escape of ammonia, they may be put into such a form as will allow them to be transported, even to great distances." He acknowledged the efforts in many European cities to manufacture fertilizers from human waste but found that "the manner in which this is done is the most injudicious which could be conceived."[30]

This manufactured manure was known as *poudrette*, a French term meaning "crumbs" or "powder." It was made from the contents of "dry closets" (as opposed to "water closets") that were hauled to the outskirts of cities and mixed with various additives such as ashes, peat, gypsum, clay, or lime. Depending on the admixtures and the process for incorporating them, the results were more or less valuable as a fertilizer. The main problem was the loss of nitrogen. Liebig found that "the greatest part of their urea is converted into carbonate of ammonia . . . and the vegetable matters contained in them putrefy; all their sulphates are decomposed. . . . The mass when dried by exposure to the air has lost more than half of the nitrogen, which the excrements originally contained."[31]

Liebig recommended "calcined mud and finely divided charcoal" as additives that would effectively retain the nitrogen.[32] He advised manufacturer James Manning on his patented formulation that included "soot, wood charcoal, seaweed charcoal, animal charcoal, or phosphate of lime, sulphates of manganese and of iron."[33]

But the industry was without regulation or standards. Ingredients might include coal tar, road sweepings, and the wastewater from tanyards, for instance. This material might be worse than ineffective. It might be toxic. Poudrette soon had such a bad reputation that farmers would not buy it. Even when the poudrette was of good quality, farmers did not know it. The English town of Ely used large quantities of charcoal and gypsum in its manufactured manure, ensuring that it would retain useful quantities of nitrogen and other fertilizers, but farmers would not buy it because it had no "scent"—their way of determining the value of a fertilizer. They had no knowledge of the power of charcoal to deodorize and preserve nutrients and so they passed it by.

Jasper Wheeler Rogers and the Fall

The alternative to dry toilets was some form of sewage or "water carriage" of excrement. While cities in Europe and America had long had drainage systems for storm water and household gray water, they were not intended for human waste, and many cities had laws against disposal of excreta in sewers. But the wealthy segment of society rapidly adopted the new water closets that piped their waste directly into the sewers, and by the 1840s, cholera was epidemic in London. It all came to a head in 1858, when a hot summer produced what was

Figure 1.3. London's Great Stink of 1858. This *Punch* magazine cartoon depicts "Father Thames" introducing his offspring—diphtheria, scrofula, and cholera—to the City of London. Image by Science History Images / Alamy

afterward called "the Great Stink" and thousands of people died from cholera. London shut its doors and windows, business ground to a halt, and the government finally passed laws to overhaul the London sewerage system, though it would be decades before a plan was devised.

It was in 1858 that Irish engineer Jasper Wheeler Rogers issued a broadside titled *Facts and Fallacies of the Sewerage System of London, and Other Large Towns*.[34] Rogers called the proliferation of water closets and the lack of any action to improve the situation "the sin of the rich—sin emanating from selfish personal comfort solely, with utter forgetfulness of consequences to others."

Rogers thought he had discovered a win-win-win system. Ever since reading Liebig's report on the power of charcoal to absorb ammonia, he had been convinced that peat charcoal was the solution to the waste problem. Peat charcoal could also solve another great problem in his native Ireland: famine. In an earlier paper, "The Potato Truck System of Ireland: The Main Cause of Her Periodical Famines and of the Non-payment of Her Rents," Rogers had exposed the true nature of the Irish famine.[35] Ireland had food, but it was being sold elsewhere because in Ireland, potatoes were money. Laborers on the farming estates were paid in potatoes, not gold. When the potatoes rotted, their cash melted away. Rogers envisioned an industry that would pay cash wages for converting Ireland's peat to charcoal that could then be used to deodorize the

Figure 1.4. The furnaces for burning the turf into charcoal at Derrymullen in Kildare County, 1850. Image by The Print Collector / Alamy

filthy sewers of Dublin, London, and other towns. He would eliminate famine, clean up the sewers, and restore soil fertility with his peat charcoal.

At a lecture in 1849 attended by six hundred people, Rogers gave a demonstration, mixing peat charcoal with foul-smelling night soil. As a newspaper reported, "A few minutes before, all noses were turned away from the tin buckets in which the night-soil was brought; a few minutes after [it was mixed with peat charcoal], it was taken up in handsful and put into paper bags provided, and 'stowed away,' possibly in the same pocket with the pocket handkerchief." A panel of appointed judges then signed a certificate: "We the undersigned . . . do unanimously decide as follows: That Mr. JW Rogers has fully proven the deodorising power of Peat Charcoal upon Human Excrement."[36]

As a result of his lobbying, the Irish Amelioration Society was established by Royal Charter in 1849, and in 1850, the Society opened a manufacturing center in Kildare County with a peat drying system and retorts to produce the peat charcoal, employing three hundred people. The plan was to open a hundred such facilities throughout Ireland. The *Illustrated London News* published an article describing the facility:

> In the centre or main building are the furnaces, thirty-six in number, arranged in rows six in each, and are composed of strong sheet-iron, of pyramidal form, with iron framework and hood to protect the upper portion of the building from the flume and vast amount of heat generated from the fiery mass within the furnaces. Here, again, the turf is still further dried upon a framework so arranged over the furnaces and across the building, when it is thrown up from below . . . by men or women.[37]

In 1851, Rogers exhibited his peat charcoal privy system at the Great Exhibition, but his effort exhausted his resources, and in 1853 Rogers found himself in debtor's prison in London, his estates and effects confiscated. Still, by 1858 he was back promoting his ideas and issuing new patents, including one for a pneumatic sanitation system that would move the contents of dry closets and mix them with peat charcoal.

Rogers's system never took hold. Perhaps the cost of manufacturing and transporting the peat charcoal to London was too high, or perhaps the convenience of the water closet was just too seductive. In 1876, the Local Government Board of London reported to Parliament with recommendations on sewage disposal methods and came down firmly on the side of hydraulic sewage systems. They concluded that "none of the manufactured manures made by manipulating town's refuse with or without chemicals, pay the contingent costs of such modes of treatment."[38]

The Toilet Bowl Economy

The interest in charcoal as a soil amendment nearly died out as agriculture became more and more reliant on cheap, easily applied chemical fertilizers. A few agronomists continued to investigate and write about soil charcoal for special uses such as plant propagation, but little more was said until early in the twenty-first century when soil scientists began investigating terra preta.

Replenishment of soil carbon was taken up as a cause by the organic movement as early as the 1940s, but most farmers were not aware of the need. The passion for recycling waste was revived in the 1970s, and is receiving more attention now as municipalities struggle with burgeoning mounds of waste, but effective systems for managing organic waste are still lacking.

Liebig's warnings about nutrient depletion have continued to resonate. His biographer William Shenstone wrote in 1895: "At present we are gradually wasting a capital which we ought to make increasingly valuable, and which no human power can restore once it has dissipated. The problem is not insoluble. It has been solved by races we are pleased to regard as almost barbarians. Till we, too, attain a solution suited to our conditions, we remain mere robbers and wastrels."[39]

The problem is still with us. Chemical fertilizers have postponed the day of reckoning, but we are much closer to the edge of disaster than we realize and have very little time left to craft the suitable solution. How different would today's cities be if Jasper Wheeler Rogers had been successful in convincing society not to crap in its drinking water? Could agriculture have avoided the chemical poisons that are making people sick and destroying wildlife? What if industry had developed under the directive to always close the materials loop by recycling? What if the toilet bowl economy had never emerged and the "flush it and forget it" model of consumption was unknown?

Liebig once started an argument with Pasteur over fermentation, asserting that vinegar was produced by chemical action, not by life-forms. When Pasteur proved that vinegar was produced by microorganisms, Liebig conceded. That's how science works. Today, we know that soil is much more than inert chemicals. It is packed full of life that is essential for the growth and health of our food plants. In a sense, we have returned to the "vital" principle of agriculture.

The end came for Liebig in 1873. On his deathbed, he arranged his final experiment. He ordered his coffin and directed that "his body should be packed in charcoal and buried in Darmstadt."

CHAPTER TWO

What Is Biochar and How Does It Work?

When biochar came on the soil science scene early in the 2000s, what made its entrance so dramatic was the vivid contrast between the rich, dark earth that could grow crops, and the native soil: highly leached, yellow-red oxisols that could barely support a weed once the overlying jungle vegetation was wiped away. In these tropical rainforests, almost all of the carbon is contained in the living biomass. When a leaf falls to the ground, it is consumed almost immediately by hungry creatures. Carbon and nutrients cycle quickly and relentlessly as things eat and get eaten. Constant rain washes soluble nutrients into the ground where microbes and roots feed. Very little is left over to be stored in soil. In contrast, a northern boreal forest builds up thick layers of peat made of partially decomposed plant matter. It is too cold and dry for life to be very active and carbon accumulates in the soil.

When I visited the Amazon with the International Biochar Initiative in 2010, we toured an organic farm growing fruits and vegetables for the city of Manaus, Brazil, 50 kilometers downriver. Walking through the rows of tomatoes, peppers, and papaya trees, I could see pottery shards scattered everywhere on the surface of the dark soil. These modern-day farmers were reaping the legacy of perhaps thousands of years of soil building by the Indigenous Amazonians. The value of the land was reflected in the property value: Land with terra preta soil sold for eight times the price of lesser farmland. We spent a day on the farm looking at different areas where the soil was more or less enriched with biochar and saw a 60-foot-tall orange tree growing in the terra preta soil.

But what makes biochar so valuable? How does it work, and how can we use it to start our own soil-building processes? Biochar is a heterogeneous and chemically complex material, and its actions in soil are difficult to tease apart and explain mechanistically.

Figure 2.1. Soil cores from a farm near Manaus, Brazil, showing the differences between the older terra preta soils rich in biochar (*left*) and the newer terra mulata soils with less biochar (*right*).

Let's start by examining the role of soil carbon in general and then drill down to a nanoscale view, looking at the chemical, electrical, physical, and biological processes that biochar seems to stimulate. While there are now multiyear field studies that show the beneficial impact of biochar on different crops, it is difficult to draw broad conclusions from these studies because they use different kinds of biochar in different types of soil with different crops.

To really understand biochar, it is perhaps more useful to look first at some general functions and properties of biochar. I will look at its behavior in compost, which is like an experimental incubator where many soil processes are accelerated. Finally, I will review a few recent studies of biochar in compost in order to explain what biochar actually does when it comes into contact with organic matter and minerals—the basic constituents of soils.

The Role of Carbon in Soil

To understand biochar, we must first appreciate the role of soil carbon. Soil carbon comes in many forms. There are two main pools of carbon: organic and inorganic. Organic forms can be further divided into recalcitrant carbon and labile carbon. Recalcitrant carbon is stubborn carbon that does not like to change its form, because it is already chemically stable. Labile carbon is mobile and changeable. Labile carbon is bioavailable (in the form of easily degraded compounds such as oils, sugars, and alcohols) and physically accessible to microbes (not bound up with minerals). The organic carbon pool includes both the living bodies and the dead, decomposing bodies of bacteria, fungi, insects, and worms, along with plant debris and manure.

Inorganic carbon consists of carbonate ions, which are typically found as salts such as calcium carbonate, and dolomite minerals, mostly in the form of rocks and sand. The interaction between mineral compounds and organic carbon produces what we like to call humus, as the minerals form complex relationships with organic molecules, protecting them from the hungry jaws of microbes.[1]

The evolution of soil shows how the soil-building process works. Before photosynthetic bacteria transformed Earth's atmosphere by filling it with oxygen, soil was nothing more than a mineral mixture of anoxic green clays. After oxygen entered the atmosphere, minerals started reacting with the oxygen, and red iron oxides appeared in the soil. Good organic, rich, productive soils developed slowly, only after algae and arthropods crawled from the sea to dry land and plants took root.[2] Life colonized land and began shedding its wasted, used-up, discarded parts onto the earth, where they formed a carbon-rich banquet that allowed new life to feed and grow, using photosynthesis to pump ever more energy into the system.

Soil building is the product of a self-reinforcing, positive feedback loop—a virtuous cycle. But soil decline is also a self-reinforcing loop (a vicious cycle) that can result in catastrophic soil loss. Most forms of agriculture tend to deplete soil carbon by reducing the amount of natural organic inputs from leaf and fruit fall as well as from woody debris, as would be found in native ecosystems. However, modern, chemical-based agriculture depletes soil carbon much more drastically. Nitrogen fertilizers combined with tillage accelerate microbial respiration, burning up soil carbon faster than it is replaced. Due to the loss of organic carbon reservoirs, many soils have become nearly lifeless substrates that must be continually fed with irrigation water, mineral nutrients, and pesticides to produce a crop. Although productive in the short term, this practice is not sustainable.

Is it possible that biochar can substitute for some of this missing soil carbon? Some of the most productive and resilient soils in the world contain significant quantities of "natural" biochar. Nature makes megatonnes of biochar through naturally occurring wildfires in forests.[3] Prairie fires can also generate a lot of biochar.[4] Tall grasses burn quick and hot; however, close to the ground where the roots start, air is excluded, so the base of the grasses will char and not burn to ash. This kind of natural charcoal is present in some of the most valuable agricultural soils in the world: the carbon-rich Chernozems of the Russian steppe and the Mollisols of the US midwestern prairie states.[5] Recently, scientists have looked more closely at the Mollisols and found that they contain charcoal that is "structurally comparable to char in the Terra Preta soils and much more abundant than previously thought (40–50 percent of organic C)."[6]

But as we know, this biochar carbon does not act the same way as organic carbon. Microbes can't eat it. So how exactly does it help soil? To really understand that, you have to drill down to the nanoscale, the scale of electrons.

Biochar Carbon at the Nanoscale

Carbon comes in many forms, and the terminology used to describe it can be confusing. Organic carbon refers to the carbon in compounds that are found in living or once-living things. Those compounds include hydrogen and oxygen in the form of hydrocarbons and carbohydrates. Inorganic carbon includes the carbonates such as limestone, and even though some life-forms use carbonates to make their shells or skeletons, these compounds are still termed "inorganic." Mineral carbon refers to carbon solids like diamond and graphite as well as to the carbon gases carbon dioxide and carbon monoxide.

There are numerous ways a mineral carbon atom can be crystallized in a solid, which leads to different physical structures called allotropes. Allotropes of

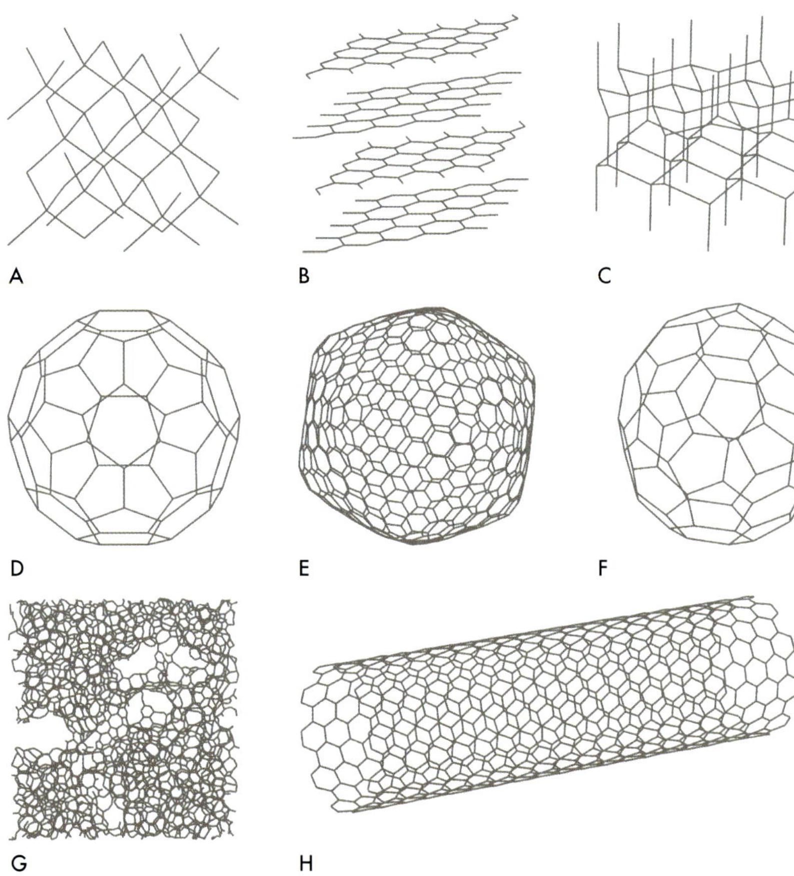

Figure 2.2. This illustration depicts eight of the allotropes (different molecular configurations) that pure carbon can take: **A.** diamond **B.** graphite **C.** lonsdaleite **D.** C60 (Buckminsterfullerene) **E.** C540 (fullerite) **F.** C70 (fullerene) **G.** amorphous carbon **H.** single-walled carbon nanotube. Image courtesy of Andel / Wikimedia Commons

mineral carbon include diamond, graphite, graphene, buckyballs, and carbon nanotubes, plus a few rare oddities.

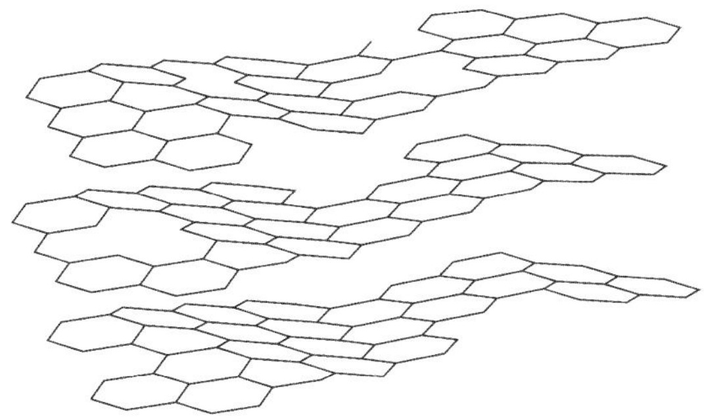

Figure 2.3. Turbostratic carbon describes carbon structures that are somewhere between amorphous (unstructured) and graphitic carbon. Image courtesy of Andel / Wikimedia Commons

You can see that many of these structures feature the six-sided carbon ring that forms graphite. Biochar is mostly graphitic, however it is not arranged in neat, crisp sheets but rather in layers and layers of discontinuous, rumpled sheets—known as turbostratic graphite crystals. This "disturbed" carbon adds tiny nanopores to the multilevel porous structure of biochar.

How biochar is made tells a lot about what the end product will look like. Biochar is made by heating biomass under the exclusion of air. Keeping air out lets the thermochemical conversion process happen without allowing oxygen in to burn up the result. The thermochemical conversion is simply the process of driving off hydrogen and oxygen that is attached to the carbon atoms and allowing the remaining carbon to link up into fused carbon rings. Take a look at the diagram of a lignin molecule (figure 2.4, page 26). Lignin is the part of a tree that gives it strength, and examining its molecular structure, you can see why it is easy to turn it into biochar. The lignin molecule has a lot of carbon rings already.

The hexagonal carbon ring is also called "aromatic" carbon (another confusing chemistry term—it does not mean that it has a strong aroma, although some carbon compounds based on hexagonal rings, like benzene, do). Fused carbon rings are very stable and microbes are not able to degrade them. The more you heat the biomass, the more of these fused carbon rings are created.[7] The more they fuse together, the more they can resist degradation. When you get sheets and sheets of them, like our turbostratic biochar sheets, they are mighty!

The fused carbon rings are also responsible for the electrical activation of biochar. Fused carbon rings form a special bond with each other called pi bonding that makes them very strong and difficult to break apart. In this form of bonding, the orbitals are out of the plane of the molecule, and the electrons can interact with each other freely and become delocalized. This means that instead of being tied to one atom of carbon, each electron is shared by all six in the ring. There are not enough electrons to form double bonds on all the carbon atoms, but the "extra" electrons strengthen all of the bonds on the ring equally. This resonant pi bonding also produces electrical properties that are used in engineered carbon materials such as graphene sheets and carbon nanotubes.

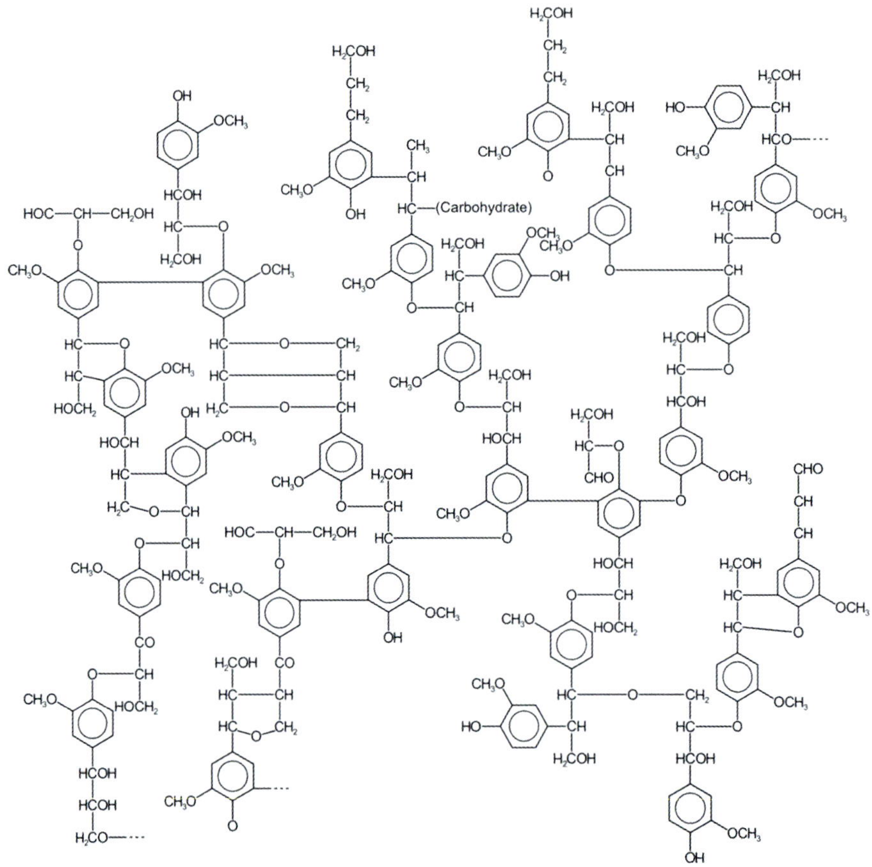

Figure 2.4. A lignin molecule, showing many carbon rings underpinning the structure. Image courtesy of Karol Głąb / Wikimedia Commons

Chemists in the nineteenth century who were working out the basics of today's science of chemistry tried to visualize the structures of atoms and molecules based on what they knew of chemical bonds. Carbon, being somewhat of a shape shifter, was difficult to visualize in compounds. Chemist August Kekule was working on benzene, a simple carbon compound composed of just carbon and hydrogen. At first, it looked like a

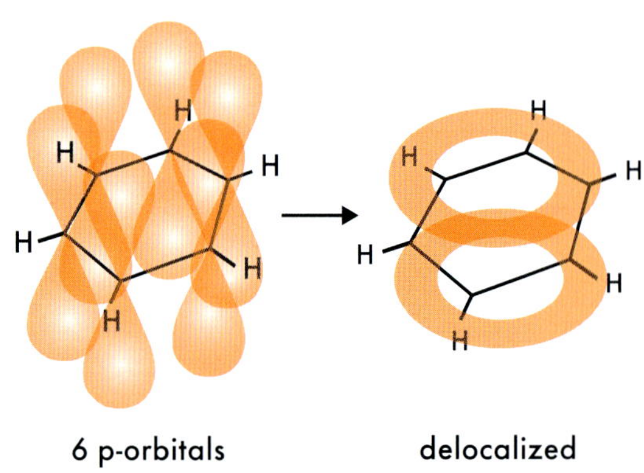

Figure 2.5. Carbon atoms arranged in a benzene ring share electrons in two planes, which contributes to the electrical properties of graphene. Image courtesy of Vladsinger / Wikimedia Commons

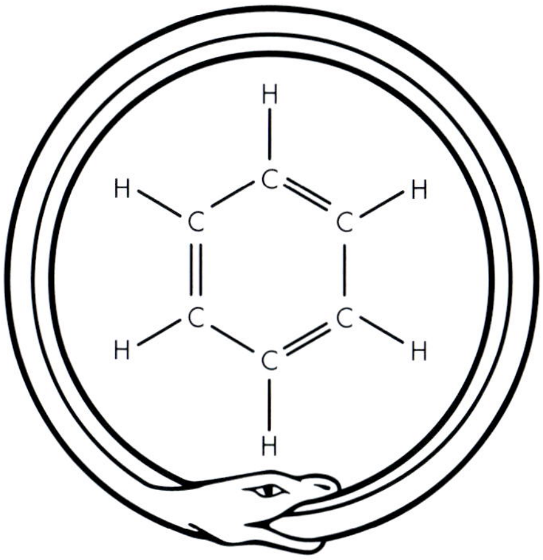

Figure 2.6. The ouroboros represents the life force and its power of regeneration. Image courtesy of Haltopub / Wikimedia Commons

sausage in his imagination, but he told his fellows that he was dreaming in front of his fireplace when the image of a whirling snake formed in his mind. The snake was an ouroboros, the mythological self-devouring snake that swallows its own tail. This is not a bad metaphor for the hexagonal carbon ring with delocalized electrons spinning around the ring.

The ouroboros is a symbol of the eternal, returning cycles of life, death, and rebirth. It is a fitting image not just for the role of biochar but for the function of the carbon ring in general—some of the most abundant carbon compounds in nature, including lignin and cellulose, are based on aromatic carbon. The electrical properties of the carbon ring are key to their function.

Electrically active fused carbon rings support "redox," or oxidation and reduction reactions that are important to soil biochemistry, by acting as both a source and sink of electrons. In soils, microorganisms use aromatic carbon both as an electron donor and as an electron acceptor during metabolic chemical reactions.[8] Biochar seems to not only serve as an electron buffer for redox reactions but it also helps bacteria swap electrons among themselves, improving their metabolic efficiency as a microbial community in something called direct interspecies electron transfer, or DIET.[9] This property is one of the reasons why biochar supports robust and highly diverse microbial communities.

The Electric Carbon Sponge

Meanwhile, at the macro- and microscales, biochar mostly looks like a blackened, shrunken version of the original biomass. This is easy to see in chunks of wood charcoal where the vascular structure—the pores and channels for xylem and phloem—are still visible. The char has a skeletal structure that looks like a carbon sponge.

Porosity comes in many scales, from the relatively large vascular and cellular structures preserved from the original biomass to the nanopores formed by tiny molecular dislocations. The amount of porosity depends mostly on the feedstock material, particle size, and the heat treatment temperature (HTT).[10] Temperature determines how much of the volatiles (hydrogen- and oxygen-containing

Figure 2.7. Biochar as a shrunken, blackened version of the original wood feedstock.

compounds) will be driven off and how much pure carbon graphite will be formed. Generally, porosity increases as more volatiles are driven off, clearing the pores, although they can re-clog if vapors are incompletely driven off and allowed to condense on the forming biochar surfaces. Then, at temperatures approaching 1000°C, pores begin to collapse or melt into smooth graphite sheets. For this reason, HTT is a key variable to know when specifying a biochar for a particular purpose. Figure 2.8 illustrates some of the differences between biochar and graphite.

Porosity will also depend on the feedstock, with high-ash feedstocks like grass reacting quite differently to heating than low-ash feedstocks like wood or bamboo. For wood feedstocks, porosity typically peaks at HTT of about 750°C.[11] Fortunately for us, 600 to 750°C is the usual temperature of formation of the biochar made in our low-tech, flame-carbonizing methods, discussed in the next chapter.

Figure 2.8. The skeletal structure of biochar looks like a carbon sponge. SEM images of biochar and graphite: **A.** Images of biochar structure and surface; **B.** Images of graphite structure and surface. Images courtesy of Ran Zhang, Haoxiang Wang, Jie Ji, and Hainian Wang / Wikimedia Commons

Porosity is directly related to surface area—the more pores, the more surface area surrounds the pores. Activated carbon is made from charcoal (or from mineral coal) that is treated to make even more pores and surface area. Imagine this: A teaspoon of activated carbon can have as much as a football field of surface area packed into all those tiny pores! Activated carbon is used in water filters to catch metals and organic contaminants on all that surface area. The charcoal we make in flame-cap kilns is almost as porous as activated carbon and can be a good substitute for it in many applications.

With its pores and its electrical charges, biochar is capable of both absorption and adsorption. Absorption (AB-sorption) is a function of pore volume. The larger pores take in water, air, and soluble nutrients like a normal sponge. Adsorption (AD-sorption) depends on surface area and charge. The surfaces of biochar adsorb materials by electro-chemical bonds, working like an electric sponge.

Surfaces are where most of the action is. Surfaces host the whirling electrons and electrical charges that hold on to all the goodies that plants want:

- Cations
- Ammonium
- Calcium
- Magnesium
- Organic compounds
- Organic functional groups
- Humic acid
- Anions
- Phosphate
- Nitrate
- Sulfate
- Minerals
- Metals
- Clay

A Well-Aged Cheese

Biochar is not soil. The electric carbon sponge is only an ingredient in the mineral and organic stew that makes up soil. The dish is usually potluck, composed of whatever the local geology and biology provide. However, the terra preta soils are different. Created over centuries by people living on densely settled high bluffs along the Amazon River, scientists hypothesize that the ingredients included charcoal, ash, food scraps, and human excrement, but how they actually combined to form terra preta is unknown. Explaining the formation of terra preta is like determining the recipe for a fine cheese. You can analyze all the ingredients and still have not the faintest idea how to make one if you don't learn it from the artisans.

One thing that is becoming obvious after two decades of biochar scientific research and the results from multiyear field trials is that, just like with a good cheese, the time dimension is critical. From the moment that biochar is pulled from the kiln, its surfaces begin to oxidize and form new compounds. These changes result in different molecules becoming attached to the surface, called "functional groups," composed primarily of oxygen, hydrogen, and carbon. These functional groups are able to bond with nutrients and minerals, while the fused carbon rings support redox reactions and shuttle electrons around the microbial community attached to biochar surfaces, enhancing microbial metabolism and the cycling of nutrients. The end result of this ferment could be any one of many "terroir"-distinct terra preta flavors, depending on what kind of soil, organic matter, minerals, water, and life-forms come into contact with the biochar, and how long it is allowed to ripen. But, if you sample the cheese before it is mature, it's just sour milk.

Raw biochar placed in soils before it has a chance to collect a charge of nutrients can actually reduce crop yields because 1) it reduces the availability of plant nutrients by binding and immobilizing them or 2) it may add volatile organic compounds (oils and tars that have recondensed in the biochar pores) that feed a bloom of microbes that use up nitrogen in the soil, depriving plants. These problems are easily corrected by adding fertilizer to the biochar application to compensate for this effect. Once the easily degradable carbon fraction is used up, biochar enters a new phase—a deep time dimension where its carbon matrix is stable for hundreds to thousands of years and may become the core of humic substances that crystallize around the fine biochar particles.[12]

In fact, biochar, whether naturally created or man-made, may be the base of many humic materials found in soils.[13] Very little humus naturally forms in tropical soils, where high temperatures and moisture accelerate microbial decomposition, yet terra preta soils have a high content of humus. To understand why, researchers added new organic matter to both a terra preta soil and an adjacent, poor natural soil.[14] They found that more of the organic matter was retained as stable humus in the terra preta soil, with less emission of carbon dioxide, indicating more efficient metabolism by the microbial community. A combination of factors may lead to this result. Biochar surfaces adsorb labile, easily degradable carbon and retain it in compounds with minerals, supporting at the same time a large microbial community that potentially makes more efficient use of organic debris containing carbon and other nutrients. The existence of this mechanism raises the possibility that terra preta soils are thus able to accumulate additional carbon more efficiently than adjacent soils.

Our current state of knowledge about biochar is now informed by the results of the first truly long-term studies of biochar in agricultural soils. A study published in 2020 titled "Soil Carbon Increased by Twice the Amount of Biochar Carbon Applied after 6 Years" confirms that biochar can be a catalyst for continual accumulation of carbon in soil.[15] The mechanisms for doing that are detailed in another study titled "Microspectroscopic Visualization of How Biochar Lifts the Soil Organic Carbon Ceiling."[16] This work describes

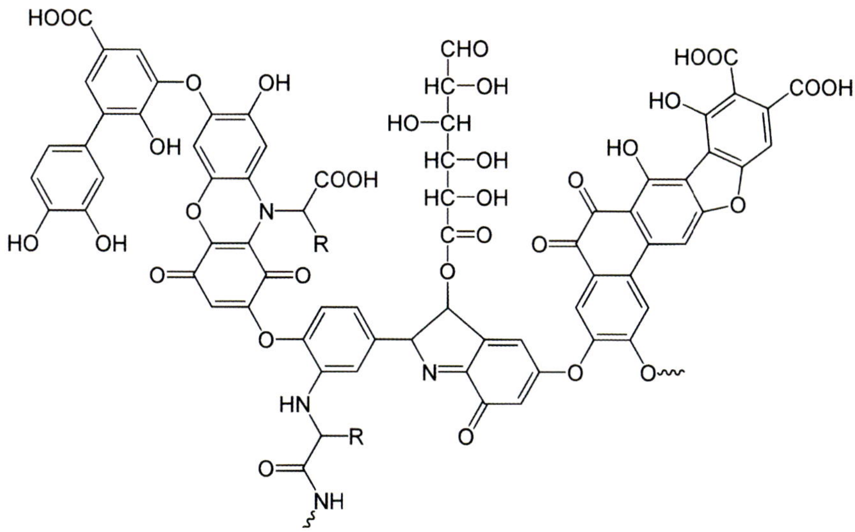

Figure 2.9. An example of the potential structure of a humic acid—another compound replete with fused carbon rings. Image courtesy of Yikrazuul / Wikimedia Commons

how the biochar electric carbon sponge holds on to microbe poop, dead cells, and perhaps most importantly, root exudates. These exudates are sugars that plants drip from their roots to feed the beneficial microbes that help feed them, especially mycorrhizal fungi. Biochar stores it all up, saving it for a rainy day and increasing the carbon and nutrient content of soil.

Kickstarting Compost with Biochar

If tropical soils need biochar to make humus, what about compost? Well-balanced compost, with the optimum C:N ratio, will contain lots of humus. However, if there is not enough stable carbon (from wood, straw, or other lignin sources), then the easily degradable sugars, fats, and proteins will be completely consumed by microbes, leaving very little substrate behind. This is what happens in tropical forests where heat, moisture, and high microbial activity quickly decompose plant litter, allowing very little soil to form.

If you look at a list of things biochar is supposed to do in soil, you'll find it is very similar to lists you see for compost. Both biochar and compost are said to provide these benefits, taken from various claims made by biochar and compost manufacturers:

- Improves tilth and reduces soil bulk density
- Increases soil water-holding capacity

- Increases cation exchange capacity (CEC)—the ability to hold on to and transfer nutrient cations: ammonium, calcium, magnesium, and potassium
- Improves fertilizer utilization, by reducing leaching from the root zone
- Retains minerals in plant available form
- Supports soil microbial life and biodiversity
- Helps plants resist diseases and pathogens
- Helps plants grow better in high salt situations

If compost really can do all these things, why do we need biochar? The answer is twofold: First, unlike biochar, compost is quickly broken down by microbial action in soil over months to, at most, years, depending primarily on climate. Recently, I called a California agriculture extension agent with a question about adding compost to fields to improve water-holding capacity. I was told that because of the hot climate, at least two applications a year are needed to maintain enough soil organic matter to make a difference in water-holding capacity, because the compost is oxidized so quickly.

Second, biochar has important synergistic effects when added to compost. Researchers find that biochar makes faster, more nutrient-rich, more biologically diverse, and more humified and stable compost. Biochar and compost can each provide what the other lacks. Biochar needs nutrients and compost needs stable carbon. Here are some of the benefits we can see from adding biochar to the composting process.

Biochar keeps compost moist and aerated, promoting increased biological activity and reduced emissions. The composting process is governed by various physical parameters that are subject to alteration by the addition of biochar materials as bulking agents. Some of the parameters that most affect compost are aeration, moisture content, temperature, bulk density, pH, electron buffering, and the sorptive capacity of bulking agents. Water and air are held in biochar pore spaces and in the spaces between particles. Moisture is also the vehicle for bringing dissolved organic carbon, nitrogen, and other plant nutritive compounds into contact with biochar surfaces where they can be captured. Biochar's stable carbon matrix accepts electrons from decomposing organic compounds, buffering electric charges that might otherwise be responsible for the production of smelly hydrogen sulfides and greenhouse gases like methane. A review study estimated that adding 5 to 10 percent biochar to compost would reduce methane emissions by greater than 80 percent, and ammonia emissions up to 98 percent.[17]

Biochar increases nitrogen retention. When nitrogen-containing biomass materials decay, they can release large amounts of ammonium. Ammonium is generated by microbial processes and nutrient cascades that convert nitrogen from organic forms found mainly in proteins and nucleic acids into mineral forms (ammonium, nitrate, and nitrite). These mineral forms can intermittently be converted by nitrifying and denitrifying microbes into gaseous emissions that include nitrogen gas (N_2), nitrous oxide (N_2O), and other reactive nitrogen gases (amines and indoles). These biological processes are complex and dependent on many factors, including soil type and pH, but overall, biochar tends to promote retention of nitrogen in soil and also the nitrification or conversion of ammonium to more stable nitrate forms of nitrogen.[18]

Biochar improves compost maturity and humic content. Several studies have looked at the effects of biochar on the timing and results of compost maturation and found that adding biochar to compost reduced the amount of dissolved organic carbon (labile carbon) in mature compost while increasing the fraction of more stable humic materials (stable carbon). Following the addition of 2 percent biochar to compost, researchers recorded a 10 percent increase in carbon captured by humic substance extraction and a 30 percent decrease of water-soluble, easily degradable carbon.[19] They also found an increase of fungal species diversity in the mature biochar compost as compared to the control, and proposed that these fungi were responsible for the increased humification. Another study found that sewage sludge composted with wood biochar had up to 30 percent more humic substances than the control.[20] One measure of compost maturity is seed germination rate. Immature compost will inhibit germination, but compost with biochar is much friendlier to seeds. You can easily test this for yourself, as you will see.

Biochar compost improves plant growth. Biochar seems to improve the composting process, but what do the plants think? Many researchers have experimented with various combinations of compost and biochar added as separate amendments. These studies find most often that plant growth is enhanced when biochar is added to soil along with compost. A 2013 study in Germany looked instead at biochar composted together with other materials.[21] They tested six different amounts of biochar in compost, from 0 to 50 percent by weight, and also three different application rates of each compost type. Using oats in greenhouse pots, they found that plant growth increased with increasing application rates of each type of biochar compost. This result is not surprising since the amount of deliverable nutrients was increased, at least by the compost fraction. However, they also discovered that plant growth was increased as the amount of biochar in the compost increased. The biochar may have either improved nutrient retention during the composting process with subsequent enhancement of nutrient delivery to plants, or promoted plant growth through some other mechanism. Regardless of the actual mechanisms, the researchers confirmed that synergistic effects can be achieved by composting with biochar.

How Can We Put Biochar to Work in Soils?

One of the basic principles of good compost production is that the wider the variety of materials you use, the better the compost. Bruno Glaser and his colleagues, who have studied both the terra preta soils and biochar in compost, have proposed a modern material flow management that combines multiple waste streams from agriculture and food processing for producing high quality compost with maximum carbon sequestration and nutrient balance. Much of this "waste" is completely wasted today, with no attempt to capture any value from it. A conservation approach is essential to help meet the growing need for replenishing carbon, nutrients, and beneficial microbes in agricultural soil.[22]

An ideal biochar compost system would be based on a speculative reconstruction of the terra preta soils. According to this model, these areas began as garbage dumps where accumulation of food wastes, ashes, and manure were deposited. However, as populations grew, it is possible that they began to realize that the waste sites were developing into very fertile and productive areas. They may have begun to deliberately manage the material flows of plant biomass, mammal and fish bones, ash, biochar, and human excreta that likely resulted in the terra preta soils we see today.

For maximum conservation of resources, it is important to remember another principle: Use the less degradable carbon sources like biochar to help preserve the more easily degradable but nutrient-laden sources like manure and food waste. By the time food waste makes it to a municipal composting facility, it has already begun to putrefy and lose nutrients. If biochar was added as the waste was collected, it could solve many problems in food-waste composting. I believe there is much exciting work ahead to determine optimum recipes for biochar-based organic composts and ferments, exploring the effects of different kinds of biochar in combination with other compost ingredients.

From past and ongoing research, we realize that biochar has numerous possible mechanisms for its action in soils that can occur on a variety of different scales. But if the results from recent biochar compost research prove to be consistent, we now have the beginnings of a recipe book for biochar-enhanced super compost that can kickstart the process of returning carbon and nutrients to soils today. Read on for a collection of my best recipes for producing composts and fertilizers with biochar. Hopefully all of this background will help prepare you to experiment with your own recipes, using what's available to you.

CHAPTER THREE

How to Make Your Own Biochar

A big part of the fun of biochar is making it. I never knew I was a pyromaniac until I got into biochar. But who doesn't love a thrifty campfire, a blazing bonfire, or a mouth-watering barbecue? Break out the marshmallows and I will tell you a story. In the old days, the hearth was the center of the home. Images appeared in the dancing flames, fueling the imagination of poets and musicians, the source of culture and communal life. Fire also powers our social engines, and that's a lot of the attraction of the biochar life for me.

Making your own biochar can be as simple as that campfire. If you have a match and some sticks, you can do it. However, that doesn't mean you don't have a lot to learn about how to do it safely, efficiently, and cleanly. I give a lot of workshops and trainings on how to make biochar—eighty or more in the last ten years, and I always start by telling people how NOT to make biochar.

How NOT to Make Biochar

When you tell people that biochar is just a form of charcoal and we can make it ourselves, you might get a bad reaction. Charcoal making can be a dirty business and the image of a charcoal maker is of someone whose face and hands are covered with soot, with billowing smoke in the background pumping out of a dirt mound. Indeed, this is the traditional method of making fuel charcoal. We don't want to do that.

There are three basic ways of making your own charcoal: a traditional mound kiln, a retort kiln, or a gasifier. The first is best avoided, the second can be very good if it is well managed, and the third is the most versatile.

The traditional way of making charcoal is to pile wood in a pit and cover it with a dirt mound. You ignite one part of the pile and let the fire slowly smolder its way through the fuel. This is called an autothermal process because the feedstock for the charcoal is also the fuel providing the heat source for the thermochemical conversion. There's not much oxygen, so there's not much flame and the pile pumps smoke out of an exhaust hole for days. It's a slow process. Some of the smoke recondenses in the pores of the charcoal, yielding a heavy, tarry charcoal that is good for its intended purpose: cooking fuel. However, this is NOT stuff you want to put in soil.

Figure 3.1. Improved FAO kilns may be more efficient producers of charcoal, but they still emit a lot of smoke. Photo by Rob Read / Alamy

Because this process is so inefficient, it has led to massive deforestation in Africa where charcoal is an essential cooking fuel. The UN Food and Agriculture Organization (FAO) designed a metal ring kiln that is more efficient so that more charcoal can be produced from the same amount of wood fuel, but it works basically the same way, and it's still a very polluting process.

The Retort

Compared to the traditional charcoal kiln, a retort kiln is a better way to make charcoal. A retort is just a vessel with a hole in it to allow gas to escape as the biomass is heated. A retort is not an autothermal process because the biomass is heated indirectly, and the heat has to transfer from a furnace or other heat source through a metal wall to heat the biomass inside. Indirect heating means the heat has to not only go through a metal wall, it also needs to penetrate deep into the retort chamber and into the center of the piece of wood that is in the center of the chamber. That can take a lot of heat and a lot of time, so in practical terms, a batch process retort will most likely not have more volume than a cubic meter or so.

The first kiln I made was a retort kiln out of a 55-gallon drum with a recirculating pipe to burn the pyrolysis gases under the drum, like a gas burner in an oven. The process was fairly time consuming. You had to split up your wood into small pieces and carefully stack it in the drum with air space between each piece, then brick up the outer enclosure and build a fire underneath. It took at least an hour of heating for the smoke to start coming out and the first

Figure 3.2. **A.** Feedstock must be properly sized and carefully loaded into the retort. **B.** The first gas that emerges is mostly steam and too wet to burn. **C.** When the hydrocarbon-rich gas comes out, you can stop stoking the fire. **D.** The yield is about 30 gallons (115 L) of biochar.

20 minutes or so was wet smoke that did not burn very well. But by the end of the process, the wood gas was "blowin' and goin'" as my blacksmith friend (and biochar pioneer) Peter Hirst would say. You could stop feeding the fire then, as the wood gas alone continued to heat the drum. At the end of the five-hour session, we had some nice biochar, but only enough to fill one wheelbarrow.

New England Biochar makes a large retort kiln with some special advantages. Not only is the gas recirculated to add heat to the process, but the kiln can take the first charge of wet gas off the fuel (that won't burn) and condense it into an interesting chemical called wood vinegar. Wood vinegar is the same thing as "liquid smoke"—you might have seen that at the grocery store. It adds that great barbecue flavor to food. Wood vinegar is very acidic, but in small amounts, it is something that plants appreciate. We'll talk more about that later on in chapter 6.

Batch retort kilns were used in the nineteenth century to make wood tar, turpentine, and other chemicals for industry, but those were soon supplanted by petroleum products. Continuous retort kilns are another beast, and several current biochar companies, such as Iowa-based ARTi, use this technology. A big advantage of a continuous retort kiln is that you can pyrolyze a wide variety of materials, including wood chips, chopped-up straw or corn stover, nut shells, and even manure and sewage sludge. Energy can be recovered from the process for heat. The ARTi unit is modular and expandable, and it can also produce wood vinegar. A continuous retort like this would be an excellent choice for a large farm or small town to become more self-sufficient in energy and fertilizer production.

Continuous retort kilns are highly engineered and come with a price tag that small farms would have trouble affording. For the small-to-medium-size garden, farm, or community, we need a different solution. That solution is gasification. Gasification is a term that covers a huge variety of technologies, ranging from large power plants to small cookstoves. Gasification also includes the primary technique I want to tell you about here: flame carbonization. Before we delve into gasification, let's talk more about pyrolysis and fire.

Basic Burn Science

There are many different ways to make biochar. The general term for the process is pyrolysis, from the Greek words for fire (*pyro*) and separation (*lysis*). Pyrolysis is separation by fire. Wood and other kinds of biomass are mostly made of the elements carbon, hydrogen, and oxygen. Pyrolysis separates the carbon from the other elements by fire, producing volatile hydrocarbon gases and leaving behind solid carbon. In the process, the carbon is also transformed into a more stable state, known as charcoal or biochar. Biochar differs from charcoal mainly in how it is used, so it needs to be fit for its purpose. We will talk about the important properties of charcoal used as biochar in chapter 5.

Fire is a thermochemical reaction between oxygen, hydrogen, and carbon that releases energy. The fire triangle shows us the three elements that are needed to start and sustain a fire. This is important to keep in mind, especially when we get to the end of the biochar-making process, which involves quenching the fire at the right stage to preserve the biochar. Our objective is "combustus interruptus." We don't want the thermochemical reaction to complete and produce ash as the only solid product. We want to stop the reaction at the right time to preserve the charcoal.

Figure 3.3. The fire triangle shows the three essential components for starting and maintaining a fire. To put a fire out, remove any one of the three elements. Image courtesy of Gustavb / Wikimedia Commons

Pyrolysis is a stage in the process of combustion. When enough heat is applied, the chemical bonds in biomass break. The more heat, the more bonds break. The elements then recombine into new molecules: the products of pyrolysis. The products of pyrolysis are charcoal and the gases that are released. The gases will burn if there is enough heat and oxygen present. Otherwise they will cool and form particulates that create smoke. Some types of pyrolysis equipment can also capture these gases and condense them into hydrocarbon liquids that can be burned as fuel or refined into useful chemicals.

As the volatile gases are released in a fire, they burn in a flame, providing heat for the process. Meanwhile, much of the carbon remains, and at a high temperature, it recombines in a new form. This is the thermochemical conversion of carbon into biochar. Once the hydrogen and oxygen leave the area, the dangling carbon bonds go looking for each other and they link up to form large sheets of hexagonal carbon rings. These bonds are hard to break; this is the secret of biochar's longevity in soil, and the reason why it can serve as a long-term carbon sequestration pathway.

But pyrolysis is only the first part of the reaction we call fire. The final part of the reaction occurs when the char itself burns, releasing carbon dioxide and carbon monoxide and leaving behind the mineral ash. Our challenge in making biochar is to get heat to convert the biomass to char, while excluding oxygen at the point where the char starts to oxidize. We call this process carbonization.

Now you may be wondering: How do you heat wood without oxygen? How would you do this at home? You would put it in a pan with a lid (basically the same as the retort discussed above), or in an externally heated oven. If I put a tray of cookies in my oven and turn up the dial to 500°F (260°C) for an hour, I will end up with little pucks of char, but the cookies will never burst into flame in the oven because there is not enough oxygen. I will, however, get a lot of smoke, and if I open the oven door and light a match, the smoke could ignite and flame.

Combustion needs oxygen. Carbonization needs the absence of oxygen. There are two main ways we have of keeping oxygen out of the process. One is by using an externally heated oven or a retort. The other method is a bit counterintuitive—we can exclude oxygen from the biomass feedstock by using the flame that burns the gases produced in the carbonization process.

My fellow enthusiasts and I have designed several types of biochar kilns we call flame-cap kilns. The flame consumes all the nearby oxygen and because it sits right on top of the burning biomass, it excludes air from the wood itself. The flame burns the gases, not the wood, and that's an important distinction.

The key thing to remember is that wood burns in three stages. The first stage is dehydration—releasing the water held in the wood. The second stage, called devolatilization, occurs when heat releases the hydrocarbon gases. If oxygen is present, those gases will ignite. If not, then the gases cool and form what we call smoke.

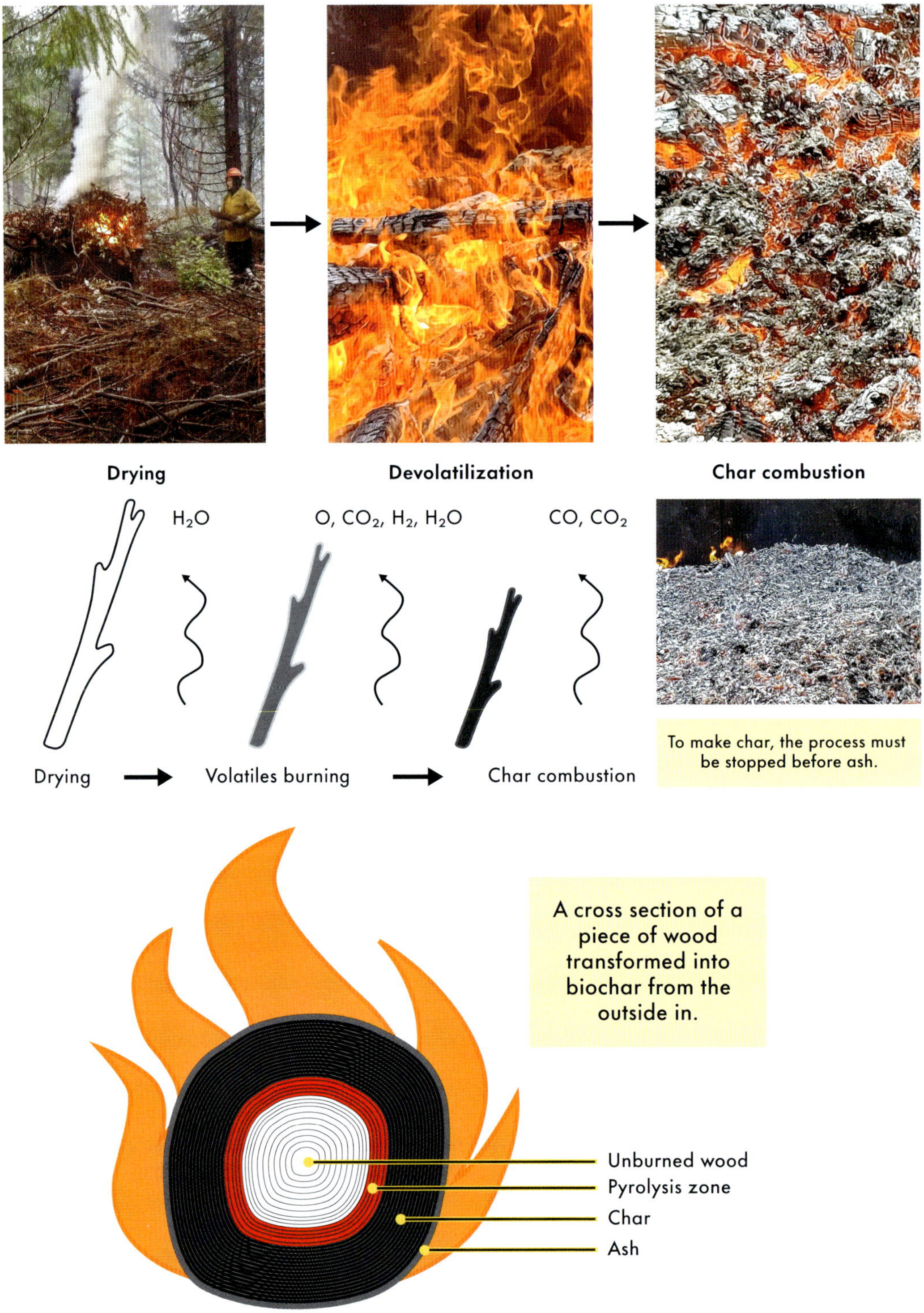

Figure 3.4. The process of flame carbonization. Biomass burns in three stages. To make char, stop the process before it goes to ash.

After the gases are all released, what is left is mostly pure carbon, or charcoal. If oxygen is present, the charcoal will continue to burn, but as most of us know from our barbecue experiences, charcoal only glows with at most a pale yellow-red aura, not the robust, roaring flames of a campfire. Charcoal is a solid fuel and only oxidizes at the surface where oxygen can reach it. Once the flames are gone, the oxidation of the remaining carbon slows down and it is easy to put it out and save the charcoal. If you do not quench it with water or snuff it out with a lid, the charcoal will continue to burn until there is nothing left but ash, which is everything mineral that will not burn.

The Flame-Cap Kiln

I first heard about the Japanese cone kiln from a paper titled "Properties of Cinders from Red Pine, Black Locust and Henon Bamboo" by Inoue et al., presented at the Asia Pacific Biochar Conference in 2011.[1] The paper showed a picture of a small metal, truncated cone called "smokeless charcoal kiln Moki M50." The kiln was 50 centimeters (20 in) in diameter and looked to be about 30 centimeters (12 in) tall. The kiln looked simple, like something I could make.

I called up my friend Kevin the welder, showed him the picture from the paper, and asked him to make me one. Once I had the kiln, though, it took me

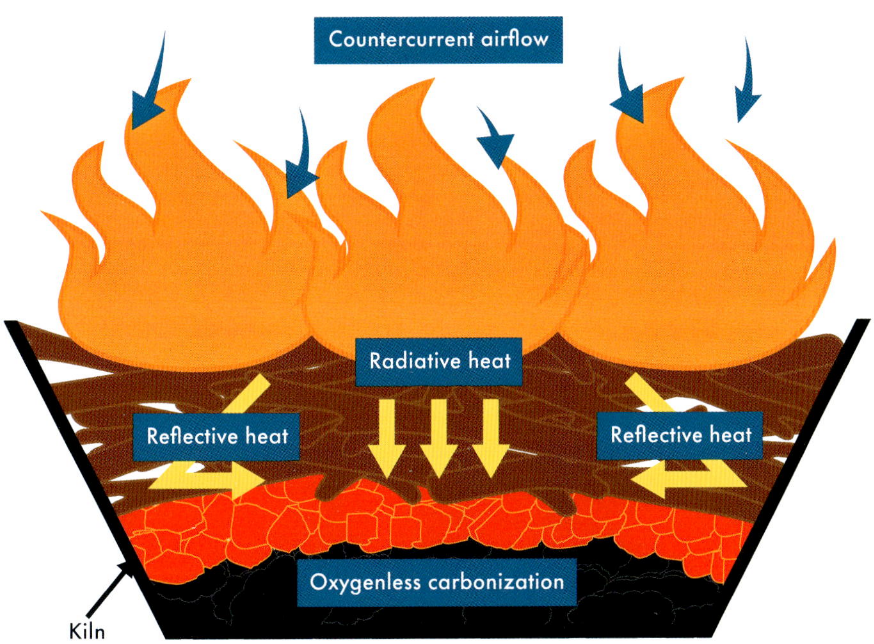

Figure 3.5. The Japanese cone kiln. We call this class of kiln a flame-cap kiln. In a flame-cap kiln, all the air comes from the top in counterflow combustion. Char that is formed at the bottom is preserved until the end, because air cannot get to it.

a few tries to figure out how to use it. All I had to go on was a poor translation of the Japanese paper. When I realized how simple and easy the whole process was, I was hooked and launched on a mission to produce more kilns.

I wanted a bigger kiln, but Kevin did not want to make another cone kiln. The first one almost killed him, he said. He's a shade tree welder and it was no

Figure 3.6. My first flame-cap kiln was a cone kiln inspired by pictures of the Japanese cone kiln.

Figure 3.7. Counterflow combustion caused the flame tips to curl down in this Oregon kiln deployed in an area burned by the 2018 Camp Fire in Paradise, California. Many homes and lives were lost in that fire. This flame shape reminded me of a phoenix, a fitting symbol for this community as they rebuild.

treat trying to bend the metal into a cone without a differential slip roll. He also did not like the waste generated by cutting a round shape out of a square sheet. So, we moved on to pyramids. That's how the Oregon kiln was born.

Figure 3.8. The Oregon kiln weighs over 200 pounds, and to get six of them stacked in my trailer, I added a small jib crane.

In 2016, I started working with the Umpqua Biochar Education Team, and we were trying to take on bigger forestry projects. We needed something that could handle big slash piles out in the woods. The Oregon kiln is a 5-foot (1.5 m) square on top and a 4-foot (1.2 m) square on the bottom with 2-foot-tall (60 cm) sides. There is a drain hole with a plug so you can flood quench it. The Oregon kiln was perfect. The plans for the Oregon kiln are available on my website, WilsonBiochar.com, and at Appropedia.org—a sustainability wiki.

We eventually made about fifty of the Oregon kilns and they are still in use and still work very well. But they are heavy and it takes four people to move one. So, I took another look at the Ring of Fire kiln that Scott McKain and Don Morrison from the UBET group had made out of corrugated roofing steel sheets bolted together. Don Morrison then made an upgrade to this unit by having a metal shop craft three panels with flanges that can be clamped or bolted together to make a ring. The beauty of the Ring of Fire is that it is lightweight and easy to transport, and you can take it apart to remove the char at the end.

Figure 3.9. The original Rings of Fire at a UBET biochar session.

In 2020, all my spring biochar workshops were suddenly cancelled and I was stuck at home, wondering what to do with myself. I took the opportunity to design a Deluxe Ring of Fire with an integrated heat shield. I had a prototype made and then recruited some bored teenagers to help me try it out. That spring we went around to all the neighbors and turned their burn piles into biochar using the best, most-efficient kiln yet. Since then, the Ring of Fire has gone through another design upgrade to make it even easier to transport and assemble. I am now manufacturing this kiln and have sold several hundred units to customers throughout the United States and Canada. You can order your own Ring of Fire® kiln at my website: RingofFire.earth.

LOVE—IT'S A BURNING THING

I love the Ring of Fire. I get a thrill every time I use it. The best part of it might be the heat shield. When operating the Oregon kiln, which has no heat shield, I was exposed to the intense radiant heat of the flame. At the end of the day, my face would be red and I would have to drink gallons of water to stay hydrated. Working around the Ring of Fire is totally different. The outside of the heat shield never gets hotter than about 350°F (180°C). That's very comfortable on a cool fall day, and it won't make you faint with heat and desiccation on a warm spring day.

The Ring of Fire with the heat shield is also the most efficient kiln I have used. It is much bigger, too, so it can take long branches without needing to cut them down and I can fill the kiln in about four hours in the right conditions, making about 3 cubic yards (2.3 m^3) of biochar. If I wanted to buy that much biochar, I would pay about $100 per cubic yard (0.8 m^3), and I would have to go get it and transport it home.

How does the Ring of Fire kiln work? The basic principle is the same for any flame-cap kiln, any kind of simple container that excludes air from the bottom and allows air to come in only from the top. Let's go through the steps, and then I'll tell you about some general design principles for making your own flame-cap kiln.

BASIC PRINCIPLES AND METHOD FOR OPERATING A FLAME-CAP KILN

The basic principle of operation is that of counterflow combustion. All combustion air comes from above. The air feeds a flame that is always maintained. Radiant heat from the flame heats the feedstock below, which emits gases that are burned in the flame. The flame consumes all of the available air so that no air is available to burn the char that forms beneath the flame. Counterflow combustion air keeps the flame low and prevents emission of embers or sparks. The flame also burns most of the smoke, making a clean burn. In summary, the following principles explain the operation of counterflow combustion in a flame-cap kiln:

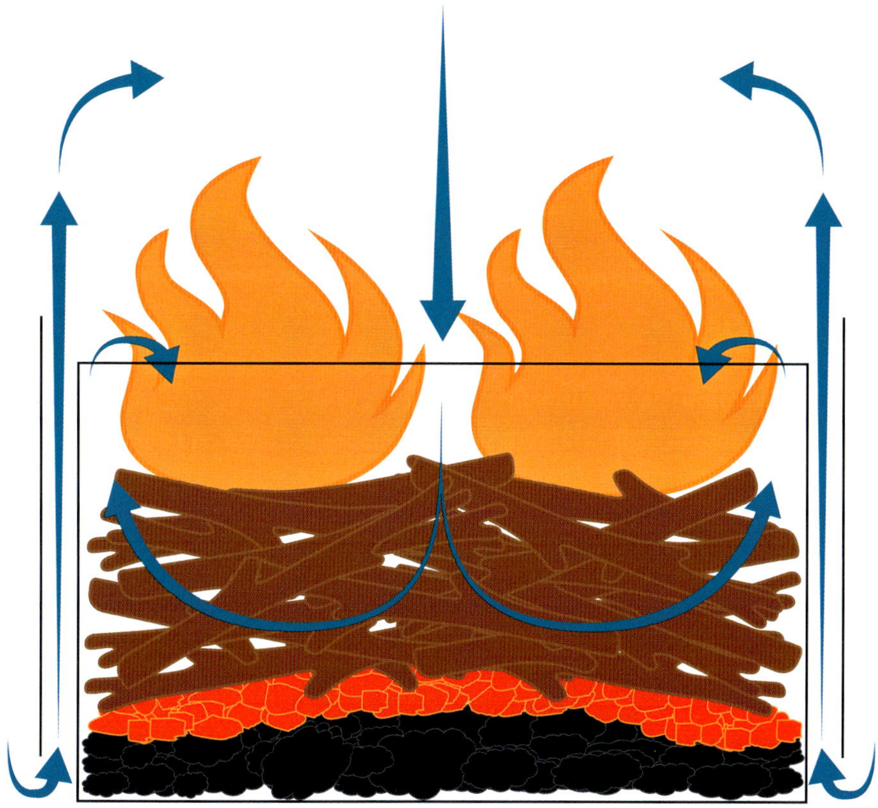

Figure 3.10. Ring of Fire kiln. Counterflow air from the top keeps embers contained and flame lengths low.

- Gas flows upward while combustion air flows downward.
- Countercurrent flow is established as burning fuel draws air downward.
- Flames stay low and close to the biomass, minimizing ember escape.
- Smoke recirculates inside the kiln and burns in the flame.
- Very little air is able to reach the unburned biochar that falls to the bottom of the kiln.
- The biochar is preserved until the end of the process when they are quenched with water.
- When a heat shield is present, it insulates the kiln against heat loss by radiation.

Now that you understand the principles behind it, here is the basic three-step method for operating a flame-cap kiln:

1. Start by building an open pile in the container, lighting it on top, and letting it burn until it collapses and coals begin to form.

2. Switch to the second stage, adding small amounts of new material on top of the coals, one layer at a time, until the container is full.
3. When the kiln is full of hot char, and the flames are gone, quench it with water to save the charcoal.

Figure 3.11. You can see how the kiln sucks smoke back into the hot zone and burns it up.

The kiln must be continually tended, adding new fuel before the charcoal turns to ash. As each new layer of fuel bursts into flame, the heat transfers by radiation into the partly charred material underneath, which continues to char, releasing gases for the flame. The flame also consumes all the air that might otherwise reach the char underneath. The combination of flame on top and the closed bottom preserves the char until it can be quenched and saved. Below are more details to help you with a successful burn:

Determine feedstock moisture. Ideally, moisture content should be between 10 percent and 25 percent. Wet wood should be tossed aside and not used until it dries out. You will burn up too much wood to create the heat needed to dry wet feedstock. However, if most of your wood is within the proper moisture range, it won't hurt to throw in a few wet or green pieces. Wood that is too dry can also cause problems: It can burn too fast and emit black particulates. Be cautious when burning extremely dry wood. A moisture meter for checking firewood costs about $30.

Choose feedstock size. Ideal size is between 1 and 4 inches (2–10 cm) thick. If the feedstock is very dry, thicker pieces can be successfully charred. Feedstock can consist of brush, branches, small logs, old lumber, mill ends, or any clean biomass material that will not pack down. Wood chips will not work well because they pack down and exclude the air needed for combustion.

Assemble the kiln (Ring of Fire). Using the connector brackets, line up all of the panels and bolt them together to form a cylinder.

Seal the kiln bottom. You MUST seal the bottom of the kiln so that no air can enter from the bottom. Use mineral dirt, clay, or sand, not flammable material. If the kiln is not properly sealed, it will not operate properly. The kiln will get too hot and the metal will warp excessively. You will also lose efficiency as more char burns down to ash. If you are using a bin kiln with a bottom, like the Oregon kiln, skip this step.

Attach the heat shield (Ring of Fire). You can attach the heat shield so it projects 4 inches (10 cm) above the inner rim, or you can attach it so it is flush with the inner rim. You might get somewhat better airflow into the kiln if you allow the outer ring to project above the inner ring, but you also expose the light-gauge heat shield to abuse by flying logs!

Fill the kiln. Best practice is to stack an initial pile of medium-size material (1 to 2 inches [2–5 cm] thick is ideal) in the kiln. This pile should fill the kiln up to slightly above the kiln rim. On top of the pile of medium-size material, place a densely packed pile of small, dry brush for kindling.

Light the material. Light the kindling and make sure the flame is spread evenly across the top. A drip torch or a propane torch are helpful but not necessary if the fuel is dry. I sometimes use an electric leaf blower to help get the fire going.

Tend the kiln. Once the initial pile burns down to glowing coals, begin adding more wood. Add a new layer of wood when each previous layer begins to show a film of white ash. Try to keep each layer of wood the same general diameter so that charring is even. Add your biggest material in the middle stages of the burn so it has time to char completely.

Always keep a strong flame on top. Let the flame be your guide as you continue to load the kiln. If you load too much densely packed material at one time, it will temporarily snuff the flame and cause smoke emissions. If this happens, back off and let the flame rise up again. If you load the kiln too slowly, the flame will die down and the kiln will begin to cool. The flame is your heat source for making the biochar, so you want it to be good and strong.

Pay attention to the loading rate. You will need to adjust the rate of loading new material depending on the feedstock characteristics. Small, dry branches will be loaded very quickly as they char completely in a short time. Large wet logs (3 to 6 inches [7–15 cm] in diameter) are loaded more slowly, but it helps to load them sequentially in several tall, open stacks so they can dry and then flame-up all at once. Otherwise, the flame dies down and never builds back up when you load wet logs too slowly, slowing the whole process down.

Finish the burn. As the kiln fills with red-hot glowing coals, make the last few layers of medium-size material to allow any larger pieces to finish charring. The charring is complete when the flames die out. Don't worry if a few bigger pieces are not fully charred. After you quench, just set them aside for the next burn.

Quench (phase 1). When the flames have died out, begin adding water in a gentle spray to the top of the kiln. Take care not to use a strong spray because it can drive air into the kiln and force a cloud of black particulates into the steam, or even drive hot embers out of the kiln.

Figure 3.12. Some important steps in the operation of the Ring of Fire kiln:
A. Assemble the kiln and build the berm. **B.** Attach the heat shield. **C.** Fill the kiln with

D

E

F

small brush and branches. **D.** Ignite the feedstock evenly across the top. **E.** Feed the kiln continually for several hours. **F.** Feed at a steady rate to maintain the flame-cap.

Figure 3.13. A. This batch is ready to quench. There will always be a few pieces not quite done, but don't wait for them all to finish. **B.** Open the kiln, spray the hot coals with water, and rake out the char to cool. **C.** Three cubic yards of biochar.

Quench (Ring of Fire). Once the kiln walls have cooled enough to touch, disassemble the heat shield and set it aside. Then disassemble the main kiln body and move it away from the hot biochar. Keep spraying water on the biochar while you rake it thin so it loses heat. Once the biochar is cool enough to touch with your bare hands, you can transfer it into bags or containers for storage.

Quench (kiln with a bottom). If using a bin kiln like the Oregon kiln, you just need to add enough water to flood the kiln. If you are short on water, make sure to stir the char well so it all gets wet and cool. Otherwise, remaining hot pockets can evaporate the water and the char will reignite.

KILN DESIGN PRINCIPLES

It is important to know that you don't need to purchase a flame-cap kiln. You can make your own, either by using the designs I describe here, or by looking around at what you already have that can be used or modified to make a kiln. Kilns can have any shape, including cylinders, cones, troughs, pyramids, rectangles, or polygons. They should have an aspect ratio of height to width that is 1:1 or less. A kiln that is too tall will have trouble getting enough air to maintain combustion.

Think about sizing the kiln to accommodate the feedstock you have. For instance, a group called Warm Heart Worldwide works with small farmers in Asia and Africa and they use a long, trough-shaped kiln for converting long pieces of bamboo into biochar.

Not everyone has a lot of woody debris to turn into biochar. Perhaps you have a suburban backyard with just a few sticks lying around. Make biochar in your kettle grill! Just close the bottom vents, make a small fire with sticks in the

Figure 3.14. You can make biochar in your kettle grill. Two small Oregon kilns (48 inches [120 cm] across and 32 inches [80 cm] across) and a kettle grill are loaded and ready to make biochar and food for workshop attendees at the Applegate Partnership in Oregon.

kettle, add more until it is full of char, and quench with water. You can even cook something on the grill before you quench it.

You can invent your own kiln design. To help you with that process, let's look at what others have done, and then I will share some general principles about heat transfer, air flow, and feedstock requirements that can help you optimize your design.

The original flame-cap kiln was the Japanese cone kiln from the Moki company. My cone kilns were modeled after it, and my pyramid kilns have a similar aspect ratio of height to width. Australian Dolph Cooke made the first cylinder-shaped kiln that I was aware of—he used a 2-meter (6.5 ft) diameter piece of heavy steel pipe that was 2 meters tall—giving a 1:1 ratio of diameter to height. He called it the Moxham kiln, named after the inventor Geoff Moxham. Sadly, Geoff died in a logging accident in 2009, but his fascinating website, Bodger's Hovel, is still online. Be very careful when working with trees, logs, and fire! Accidents can happen to even experienced woods workers.

Paul Taylor and Hans Peter Schmidt from the Ithaka Institute in Switzerland came up with the Kon Tiki kiln—a deep cone kiln that has a diameter of 150 centimeters (59 in) at the top, with a base diameter of 60 centimeters (23.5 in) and total volume of 850 liters (225 gal). The height of the kiln is 90 centimeters (35.5 in) and it sits on three legs above the ground, with a drain valve that can be fitted with a hose for quenching from below. You can find plans for constructing the Kon Tiki online at Appropedia.org. The best innovation of the Kon Tiki kiln, to my mind, was the addition of a heat shield. The heat shield on the Kon Tiki actually helps to make up one of the deficiencies I see in the design: It has a very high surface-area-to-volume ratio. Why is this important? It has to do with heat loss. A kiln that has lots of surface area loses more heat to the environment by convection and

Figure 3.15. Kon Tiki kiln with swivel mount for tipping. The heat shield is a separate attachment.

radiation. The heat shield mitigates this by reducing heat loss with an insulating layer of air.

The Big Box kiln, designed by Darren McAvoy at Utah State University, is a large bin kiln with an integrated heat shield that can be hauled on a trailer. It is big, tough, and heavy, and can be loaded using machinery. It is a bin kiln with a drain so you can flood quench it. At the end of the burn, operators use the loader to tip the kiln and dump it out, then drag it back and load it on the trailer. This kiln is perfect for large piles of biomass near a road that can serve no better purpose than to become biochar. Darren has the plans for the Big Box kiln freely available on Appropedia.org.

Figure 3.16. The Big Box kiln travels on a trailer with a loader to move the kiln and feed it with biomass. It can also be loaded by hand.

Table 3.1 shows some calculations of the surface-area-to-volume ratio for some of the kilns I have described. The surface area calculations do not include the open top of the kiln, nor the bottom of the kiln if it rests on the ground where it is insulated by dirt. Only the Kon Tiki is elevated above the ground. Surface area to volume depends on the size of the kiln as well as the shape. Big volumes have relatively less surface area to volume than small volumes do. The lesson here is: Make your kiln bigger if possible.

Other considerations for kiln shapes include the ratio of the height to the width of the kiln. A kiln that is too tall will have trouble drawing air down inside

Table 3.1. Surface-Area-to-Volume Ratio for Different Kiln Shapes and Sizes

Kiln	Shape	Top Dimensions	Bottom Dimensions	Height	Volume (V)	Surface Area (SA)	SA/V
Kon Tiki	deep cone	.75 m radius	0.3 m radius	0.9 m	.85 m^3	3.6 m^2	4.5
Moxham	tall, vertical cylinder	1.0 m radius	N/A	2.0 m	6.3 m^3	12.6 m^2	2
Ring of Fire	short, vertical cylinder	1.1 m radius	N/A	1.0 m	4.1 m^3	7.2 m^2	1.7
Big Box	rectangular box	1.8 m x 3.7 m	N/A	1.2 m	8.0 m^3	13.2 m^2	1.6

the kiln to burn material at the bottom, since all of the air has to come from above. The Moxham kiln diameter-to-height ratio of 1:1 is about as tall as you want to go. A bigger, shallower kiln will usually work better.

A useful variation on the Kon Tiki is manufactured by Australian Frank Strie. Frank has added a lot of nice bells and whistles to the Kon Tiki, renamed as Kon Tiki Tas (*Tas* for Tasmania). Frank has made the Cadillac of flame-cap kilns! First, he has built a more favorable kiln geometry (from a heat-retention standpoint) by expanding the Kon Tiki cone into a trench shape he calls the Kon Tiki Tas Stretch kiln. Each batch can produce approximately 1,700 liters (2.2 yd^3) of biochar within four to six hours. The kiln includes a cable-winch tilting mechanism for unloading the biochar. Frank likes to quench with a manure tea to immediately charge hot char with nutrients, so the Kon Tiki Tas Stretch includes a flood quenching and drainage pipe at the cone floor complete with a 50-millimeter ball valve and fast coupling. The stainless steel heat shield has an easy mounting system with hangers and clips, and the kiln is mounted on a skid frame with fork pockets so it can be moved around the farm with a tractor. There is also an optional air-injection rim tube made of stainless steel that uses a leaf blower to force air into the kiln.

Frank Strie manufactures and sells his Kon Tiki Tas Stretch kiln and a standard cone-shaped Kon Tiki Tas in Tasmania through his website TerraPretaDevelopments.com.au. He ships anywhere in Australia.

Simply digging a pit in the ground and using it as a kiln is another option. However, there are several concerns. The first is safety. How are you going to make sure that no one falls into your pit of fire? You definitely want to keep children and pets at a distance. The second concern is quenching and char removal. If the soil under the pit is free draining, it could take a lot of water

Figure 3.17. Frank Strie displays the Kon Tiki Tas Stretch kiln at a festival in Tasmania.

and a lot of stirring to get the char all cold and wet. Some people will simply cover the pit with dirt to quench, and while that can work, dirt quenching adds to the problem of getting the char out of the pit. It will be difficult to avoid mixing a lot of dirt and rocks with the char. This may not be a problem if the char is going to be mixed with soil anyway, so don't let that deter you. A better option could be to build an aboveground container out of masonry, cob, or just a berm of dirt.

Here are some items that you might have on hand already that could function as kilns: stock tanks, old water tanks cut in half, dumpster bins, defunct trailers, an old truck bed, harvest boxes, or grain bins. I have seen a large truck hood from a wreck used as a kiln. Be creative!

The Conservation Burn

Sometimes, the best kiln is no kiln at all. Perhaps you don't have money or time to acquire a kiln. Perhaps you have hundreds of burn piles out on a steep slope and no way to get a kiln up there. Sometimes you just have to burn the stuff to

Figure 3.18. Conservation burn airflow and flames.

Figure 3.19. A young worker learns about swamper burns at the On Ramp Forestry Training Camp at the Enchanted Forest.

get rid of it. We see this a lot here in the western United States where the need to remove flammable fuels from forested areas is great and funds to do so are severely limited. In these situations, the conservation burn is a great option.

In 2013, Raymond Baltar of the Sonoma Biochar Initiative sponsored a Biochar Burn Workshop with me at the Enchanted Forest property owned by my friend Brenda Patton. Peter Hirst was our main instructor and about thirty people attended. We spent three days experimenting with a hundred small burn piles scattered around the property. We tried different methods of stacking, lighting, and quenching burn piles to make biochar, and developed several techniques that we called conservation burns.

A conservation burn is simply a burn pile that is constructed and lit so that a strong flame is maintained throughout the burn, similar to a flame-cap kiln. As with a kiln burn, the flames will burn up much of the smoke, and the burn pile is quenched at the end to save the char without allowing it to burn to ash. Pile construction is the key to a successful burn. Piles need to be free of dirt and rocks, so no bulldozer piles. Small hand piles will work best, although bigger piles can be built with machinery, as long as there is air space. Optimize char production by using the right pile construction for the type of feedstock you have. If you have small branches and brush, make a simple pile up to 6 feet (1.8 m) in diameter and 4 or 5 feet (1.2–1.5 m) tall. Bigger logs can be stacked in a log-cabin-style rick by hand. A rick works well for old lumber as well. If you have mixed sizes, try arranging the larger logs around the pile of small stuff in a teepee formation. For very large piles, use an excavator to drop slash onto a pile from above so air spaces are created.

A conservation burn can also be managed like a kiln burn, by continually adding material to build up a bigger pile of char. We call this a swamper burn because workers known as "swampers" schlepp the material forward to add to the pile. The term came from the historic logging industry in the South where workers who cleared and built roads through swamps for logging crews were called swampers. Swamper burns are a common practice here in the West. Forest managers like the technique because it produces fewer burn pile scars on the land. Swamper burns are also good for burning green, freshly cut branches. Once a hot fire is going, green wood is added one branch at a time. The only difference between what forestry workers are doing already, and a char-making swamper burn, is that we put it out with water at the end to save the char.

Here are some examples to illustrate conservation burn methods to char small brush, bigger logs, and mixed sizes.

SMALL BRUSH

Small brush can produce a lot of biochar very quickly. The key is to consolidate it as it burns and put it out at the right moment. Here is one of my favorite tips: If you have some old roofing steel lying around, lay it on the ground to make a steel pad large enough to fit under your burn pile. Pile your small brush on that and burn it. The steel will reflect heat back up into the pile. It will go quickly and after you quench it the biochar will be easy to gather up and recover.

Figure 3.20. Small brush burns quickly. Use a rake to push the edges back into the flaming hot part of the fire (*left*). Three brush piles made about 1 cubic yard (0.75 m^3) of biochar in about one hour (*right*).

Figure 3.21. A. Build a rick pile and light it on the top (*top*). It does not take long for the whole thing to be engulfed in flame. You get better results if you consolidate material in the hottest part of the fire as it burns down (*bottom*). Note that all the smoke in the picture comes from logs that are not in the main part of the fire.

LOGS

To char bigger logs (those pictured in figure 3.21 are between 3 and 6 inches [7–15 cm]), make sure there is space for air between the pieces, the material is dry, and there is no dirt in the pile. Make a kindling pile on top for ignition. Keep chunking logs back into the fire as it burns. If you keep everything in the hot part of the fire, you will not see smoke. You can also pull finished char out of a still-burning pile and quench it separately while the rest of the pile continues to burn.

MIXED SIZES

Mixed-size piles are made by stacking smaller branches in the middle of the pile and then placing the larger logs in a teepee formation around the pile. The larger logs will be very effective at holding in heat and controlling airflow, so

Figure 3.22. Smaller branches are piled loosely with bigger logs laid around in a teepee (*top*). The pile retains its structure as it burns down; the teepee logs help hold it together (*bottom*).

the smaller material in the middle will char quickly and efficiently. Some of the big logs won't char completely, but that's okay—they are still valuable on the landscape for ecological benefits. Once they are completely extinguished, scatter them on the land to provide habitat for insects and fungi.

FORESTRY PILES

In the fire-prone western forests, both private and public landowners are engaged in landscape-scale thinning projects to reduce hazardous fuels that have accumulated because we have suppressed fire for so long. Historically, these forests burned regularly, clearing out the understory while sparing most of the big, old trees. The goal of these thinning projects is to reduce fire hazards to homes and property, while preparing the forests for the return of a regular program of prescribed fire.

Figure 3.23. Sequence for recovering biochar from a burn pile: **A.** Light piles near the top to achieve flaming combustion quickly; **B.** When embers begin to ash it is time to quench; **C.** Spray and rake thin to lose heat quickly and minimize water consumption; **D.** Several cubic feet of biochar can be produced from one fire. Use buckets to scatter the partially charred "bones" in the forest for ecological benefits.

Thinning projects can generate 10 to 50 piles per acre. We recently did a pile-burning job on 6 acres that had a total of 120 piles that were each about 5 feet (1.5 m) in diameter. We had a water truck that was able to extinguish about 30 piles and save a lot of char. As luck would have it, just as we finished burning all the piles, a heavy rain started to fall. We quickly went around to all the remaining piles and raked out the burning coals to a thin layer. This worked very well and the rain saved almost as much char as the piles we quenched with the water truck. There were a few piles we missed raking out, and the rain did not save much of that char. Raking it thin is the key to quenching.

At another pile-burning project, it had rained heavily the day before, so the ground was wet. We did not have water to quench all the piles, so we shoveled up the burning coals and flung them out on the wet ground. Once a hot, burning coal hits wet ground, it quickly cools and goes out. Now there are nice chunks of biochar strewn about the forest floor, waiting for microbes and fungi to move in.

Be cautious. Large-scale pile-burning projects like this need to be supervised by a trained burn boss to make sure the conditions are safe. Otherwise, practices like raking and flinging hot coals could ignite a wildfire. However, in the right conditions there is no problem.

Feedstocks

Most woody materials make good feedstocks for biochar produced using either the kiln or conservation burn pile methods. Woody feedstocks can consist of brush, branches, small logs, old lumber, mill ends, or any clean biomass material that will not pack down. Be aware that feedstocks with leaves attached can create flames high in the air that may cause burning leaves to be transported out of the kiln or pile burn. Wood chips are not good feedstocks because they pack down and exclude the air needed for combustion. Certain agricultural wastes, such as orchard prunings, corn stover, and hemp stalks, will work well in kilns and pile burns because they are branchy and don't pack down. Other organic wastes, such as corncobs or nutshells, can be processed more easily in a top-lit updraft (TLUD) gasifier. Chapter 4 describes some ways of doing that. When selecting feedstocks, there are two primary conditions to manage for: moisture and size.

FEEDSTOCK MOISTURE

Use dry feedstocks. Fuel moisture is the main factor that will affect the time to char material. Ideally, the moisture content of the bulk of the material should be between 10 and 25 percent. A moisture meter is a worthwhile investment for anyone planning to make biochar. Wood and feedstocks that are too dry (around

Figure 3.24. A moisture meter is an essential tool for making biochar. Make sure you check the sides of the log too. Wood dries from the ends.

5 percent or less) can create large amounts of black, sooty smoke because they burn too quickly and the hot gases are transported at great velocity outside the kiln where they condense into soot. Use extra caution if your feedstocks are very dry, as flame lengths will be longer than when using wetter feedstocks. Wet wood with more than 30 percent moisture should not be used because too much feedstock will have to burn to create the heat needed for the water to evaporate from the feedstock, making the process highly inefficient. Allow wet woody feedstocks to dry before burning. If they are up to 2 inches (5 cm) in diameter, allow to dry for 30 days. If they are between 2 and 6 inches (5–15 cm) in diameter, allow to dry for 60 days.

Depending on how wet the wood is, these timelines should allow the wood to fall below 30 percent moisture for making char. Make sure that you apply the moisture meter to the sides of the branch or log and not to just the ends, because the ends will be drier than the middle. Insert the pins through any bark layer and into the wood. Cutting wood to shorter lengths will help it dry faster.

Some freshly cut green fuels can burn well once the kiln is hot. These include *Ceanothus* and manzanita species, which have a high concentration of volatiles. While these green fuels may have up to 50 percent moisture, they also have additional fuel value from the volatiles, making them fairly efficient to use for char production. The extra fuel makes up for the extra heat needed to evaporate the water. This volatile fuel is lost to the atmosphere during the drying process. Some species will not burn well at all when green: Poplar holds a lot of water, especially when green. Oak is dense and low in volatiles and needs to be very dry to burn well. Keep in mind that burning wet feedstocks also results in more methane emissions. Methane is a very potent greenhouse gas, so we want to avoid producing it.

FEEDSTOCK SIZE

The ideal size of woody feedstocks is between 1 and 4 inches (2–10 cm) thick. If the feedstocks are very dry, pieces up to 8 inches (20 cm) thick can be successfully charred in a large kiln. Big pieces are less efficient because you run into the same problem you have with a retort kiln: It takes a long time to get the heat to penetrate to the center of a big log. Meanwhile you burn away most of the outer part of the log to ash.

In a kiln burn, add small brush and branches separately from larger wood pieces. If you mix feedstock sizes, the small material will be wasted because it will burn completely to ash before the bigger material has finished charring. You can process different sizes separately in the kiln by adding some larger material in the middle of the burn and finishing with smaller material on top. This allows the big stuff enough time to finish charring. Don't put it on the bottom, though, because the material on the bottom starts cooling before the batch is done, especially in a taller kiln. You will end up with a lot of unburned material in the bottom of the kiln.

Mixed sizes are sometimes the best way to char material like berry canes, vineyard and orchard prunings, hemp stalks, corn stalks, scotch broom, and reed grasses. Start a fire in the kiln with dry brush, old pallets, or other woody material that heats up quickly. Once the fire is strong and the kiln is hot, start adding the finer material. If it packs down and starts to smoke, throw in some more branchy wood to create air spaces to keep the flame hot and clean.

Safety and Tools

Always check with your local fire authorities about safe burn times and get a burn permit if they require it. Most jurisdictions will have restrictions on burning based on the season and the air quality conditions. In some places you will need to check in on the day of your burn to make sure that humidity levels

Figure 3.25. Tools of the trade: shovels and rakes, moisture meter, infrared thermometer, propane torch, electric leaf blower, metal pail for sampling char. The red-haired troll is to remind us to always keep a strong flame on top!

are high enough and there are no high winds in the forecast. If you have never managed a burn pile before, please work with someone who has and learn about safety and working with fire.

Keep your work area clean to avoid tripping hazards. Feedstock piles should not be kept right next to the kiln. Place the kiln so there are no low hanging branches overhead that could catch fire. Survey the scene while you work to make sure that no embers have landed on flammable materials and started a fire.

Make sure everyone who works with you is properly equipped for working with fire. This includes sturdy work boots, leather gloves, and nonflammable clothing like cotton or wool. Long hair should be tied back, and everyone should have plenty of water to drink. Always keep water on hand for extinguishing any embers and for quenching the kiln at the end. You also want several shovels and rakes.

Only burn clean feedstocks. Do not burn treated wood or wood with paint. Know the characteristics of the feedstocks. There are woods that release deadly chemicals when burned (oleander, for example). We have burned a lot of poison oak, however, and not noticed any ill effects from it.

You may want to acquire the following tools to help with safe and efficient biochar production:

- Wood moisture meter
- Propane torch for lighting
- Buckets for holding readily available water
- McLeod fire rake for clearing around the kiln, building berms, and raking out the char at the end
- Electric leaf blower (a big help if you are stuck trying to light damp fuel—the extra air is just what you need)

CHAPTER FOUR

Working the Flame

We have talked about flame-cap kilns and how they work, but there is more to know about flames and fire, and there are additional techniques for making biochar the low-tech way using a flame. There are also important safety considerations to be aware of when working around flames, especially if you plan to work on a large scale.

The National Wildfire Coordinating Group (NWCG) offers an excellent online course, S-190: Introduction to Wildland Fire Behavior, that covers fuels, weather, and topography, and how those elements affect fire behavior and firefighter safety. You may be able to access this training through a local volunteer firefighter program or the Prescribed Fire Training Exchanges (TREX) cooperative network. To find out more, visit firenetworks.org.

If you live in a high fire risk area like the western United States, having a basic knowledge of wildland fire behavior is valuable on many levels.

What firefighters learn about landforms, fuels, weather, and fire behavior on a large scale is based on the same principles of furnace geometry, fuels, draft, and flame characteristics that are used in designing natural-draft stoves, furnaces, and gasifiers. By covering a few of these topics, you can also get a better idea of how to design a biochar kiln or a small, low-tech gasifier for making biochar and heat, putting the flames to work and using their energy.

Flame Diversity

Flames come in two basic types having to do with the mixing of air and fuel. A pre-mixed flame combines gaseous fuel and air before the flame is ignited. This is what happens in an engine when a cylinder draws in an air/fuel mixture from a carburetor. The mixture explodes when the spark plug fires, turning the crankshaft and moving the vehicle down the road.

We are using solid fuels, so all of our flames arise from wood gas that needs to mix with air before it will burn. Usually, the flame is already present and the air and fuel burn as they diffuse and mix together. This is called a diffusion flame. Sometimes, in a structure fire, air and smoke can mix in a room ahead of the flames and will explode when the flames reach the room. This is called a smoke explosion, and fortunately, it is rare. A diffusion flame is limited by the rate at which fuel and oxidizer can combine, so it is not explosive.

A diffusion flame can have different forms, depending on air flows and gravity. The image on the right in figure 4.1 shows a candle burning in microgravity on the International Space Station. Without gravity, there is no buoyancy effect. Buoyancy gives the candle flame in Earth gravity its shape because hot air expands and is less dense, while the colder, heavier air surrounding it exerts pressure on the flame, squeezing it into the teardrop shape of a candle flame. The space flame is spherical because there is no gravity to act on the different densities. The space flame is blue because buoyant gases are not carrying unburned fuel particles up and away from the hottest part of the flame. Wax burns as gasified carbon and hydrogen molecules that emit blue light. In the Earth flame, the gasified candle wax burns very hot initially, forming the blue crescent you see at the bottom of the flame. Buoyancy pressures propel the gas away from the ignition point before the combustion reaction completes and the unburned gas condenses into particles of soot. The flame is still hot enough to burn the soot particles, but at a lower temperature that glows yellow and red, incidentally emitting light that we can read by, unlike the blue glow of the space flame.

A diffusion flame can also be either laminar or turbulent. The candle flame is laminar, meaning that the air and fuel flow in smooth layers that mix with each other mostly by molecular diffusion, which is simply the collision and reaction of the molecules of oxygen, hydrogen, and carbon.

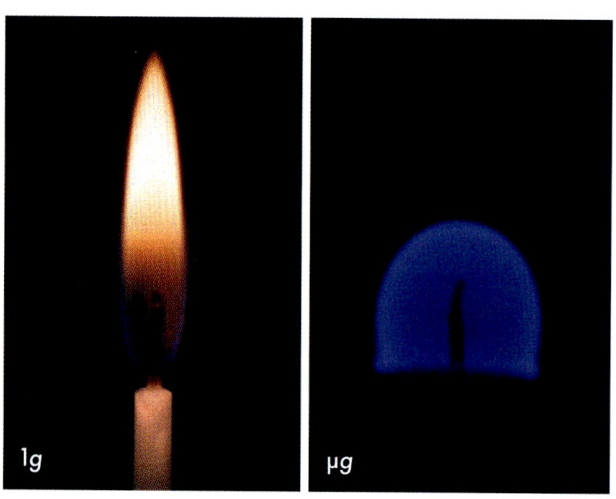

Figure 4.1. In space no one can see you by candlelight. The flame on the left burns under Earth gravity. The flame on the right burns in microgravity on the International Space Station. Photo courtesy of NASA

A turbulent flame looks a lot like turbulent water in a river with swirls and eddies. It's what you see in a campfire. The mixing of air and fuel takes place as a result of the physical transport of the gas flows. There are lots of different kinds of turbulence of course, and some kinds mix better than others. Molecular diffusion also takes place in a turbulent flame, so it can be thought of as a collection of small laminar flames called flamelets. There is much chaos involved and combustion fluid dynamics are notoriously difficult to model with equations and computers.

In a flame-cap kiln, a gasifier, or a furnace, rising hot air tends to transport fuel out of the hot zone more quickly than we like, so we can end up with soot and smoke as the gases cool and condense. Clean combustion in low-tech kilns and stoves is all about managing buoyancy. Furnace designers call this "flame shaping," and the geometry of the furnace, the placement of air and fuel inlets, radiation, and the insulation of the furnace walls all come into play. The ultimate goal is to burn fuels cleanly and completely and to transfer the heat to where you want it to go.

Counterflow Combustion

One way to tame a flame in the presence of gravity is to promote counterflow combustion. That is what we try to do with the flame-cap kiln. When air and fuel are coming from two different directions, it promotes mixing. Here are some principles for working with buoyant diffusion flames from a 1964 Fire Research Note:[1]

- A shorter flame length is better (flame length is a function of mass flow rate and fuel bed diameter).
- Open-sided piles entrain more air than enclosed piles, giving higher air mass flow rate and higher flame length.
- High flow rate flames cool faster and radiate less heat to the fuel bed.
- Piles enclosed by a surrounding cylinder have shorter, more turbulent flames due to less entrained air.

During our work in Paradise, California, following the devastating Camp Fire, we helped a landowner burn piles of dry fuels removed from near his home. We lit some of the piles as they were, and we also wrapped a Ring of Fire kiln around one. You can see the difference in how they burned in the series of pictures in figure 4.2 (page 68). The open piles burned with tall shooting flames that did not die down quickly because the mass flow rate of the air was very high. Air could come in from all sides of the pile.

In contrast, the pile in the kiln had a somewhat tall flame initially because the top part of the pile was still exposed to air. But the flame length dropped quickly, and as we continued to load the kiln with more material from other piles, the flame length stayed low.

The neighbors who came to help told me that burning in the kiln made them feel safer. Long flames are dangerous, and that community had been traumatized by flames, yet they had a lot of work to do to dispose of burn piles to keep them safe from future fires. Burning in the Ring of Fire was a better option for many reasons.

Figure 4.2. **A.** Tall shooting flame from a recently lit open burn pile. **B.** Even as the pile burns down, flame lengths stay very tall. **C.** Lighting a burn pile wrapped inside a Ring of Fire kiln. **D.** Flame lengths are short throughout the burn, allowing us to load additional material into the kiln.

A Simple Gasifier

News flash! Flames make heat! Yes, we know, we ought to recover some of that heat as useful energy while we make biochar. As discussed earlier, though, to do that at scale and produce electricity requires expensive, complicated equipment. But there are some ways, even on a small scale, to recover useful heat energy from the biochar-making process.

Certain types of feedstocks—those that come in small chunks rather than the branches and brushy material used in flame-cap kilns—can work better in a gasifier. Gasifier is a broad term and it covers a lot of different technologies, but the only type that is low-tech and easy for anyone to make is the top-lit updraft (TLUD) gasifier.

A TLUD works best with smaller, granular, or chunky fuels that don't work well in a flame-cap kiln. A small TLUD can use wood chips, nut shells, twigs, or pellets made from sawdust, rice hulls, or other crop waste. A larger TLUD—think 55-gallon-drum size—can process lumber offcuts, mill ends, corncobs, hemp stalks, sticks, and twigs. Large, chunky, or shredded wood chips can also work in a TLUD, as long as they don't pack too tightly and cut off the air flow. Many people have used the 55-gallon TLUD for their first biochar production for gardens and small farms. You will find lots of examples of the drum TLUD across the internet. Warm Heart Worldwide introduced these devices to farmers in Thailand for pyrolyzing corn cobs and crop waste, and they work quite well for that purpose.

A TLUD can serve as a cookstove. Every year, the international clean cookstove community meets at the ETHOS (Engineers in Technical Humanitarian Opportunities of Service) conference in Kirkland, Washington, to discuss progress in the goal of providing cleaner biomass cookstoves for the two billion or so people on planet Earth who only have wood and other forms of biomass fuel to cook with. With many people cooking over what are essentially campfires inside their houses, the rate of respiratory disease in women and children is especially high. TLUDs, along with rocket stoves, have emerged as a viable option for improving the situation. TLUDs can also make biochar, and advocates and inventors have created many designs and programs for providing stoves to end users. A TLUD can provide a family with clean cooking ability

Figure 4.3. Christa Roth demonstrates a TLUD made of "tincanium" at the annual ETHOS conference for clean cookstove developers.

utilizing waste like nut shells or corncobs that have no other use. If the family wants to save the biochar for use in gardens, it is easy to do. If they would rather burn the char to finish cooking dinner, they have that option, too.

In contrast to a flame-cap kiln that uses counterflow combustion air, the TLUD uses cocurrent flow along an axis, so air and wood gas flow together. A TLUD is usually a cylinder that allows air to flow through the fuel from the bottom to the top. While a TLUD can use forced air (a fan or blower) to do this, a natural-draft TLUD relies on the buoyant force to move the air. The buoyant force is also balanced against resistance to that flow by the fuel. The fuel particles need to have enough air spaces between them to allow the air to flow without too much restriction.

Airflow from the bottom is "primary air" and airflow at the top of the fuel bed is "secondary air." Primary air flows up through the unburned fuel. Then it hits what is called the "migrating pyrolysis front." This air is not enough to completely combust the fuel, but it is sufficient to supply oxygen to a hot, smoldering layer below the char, creating wood gas and forming more char.

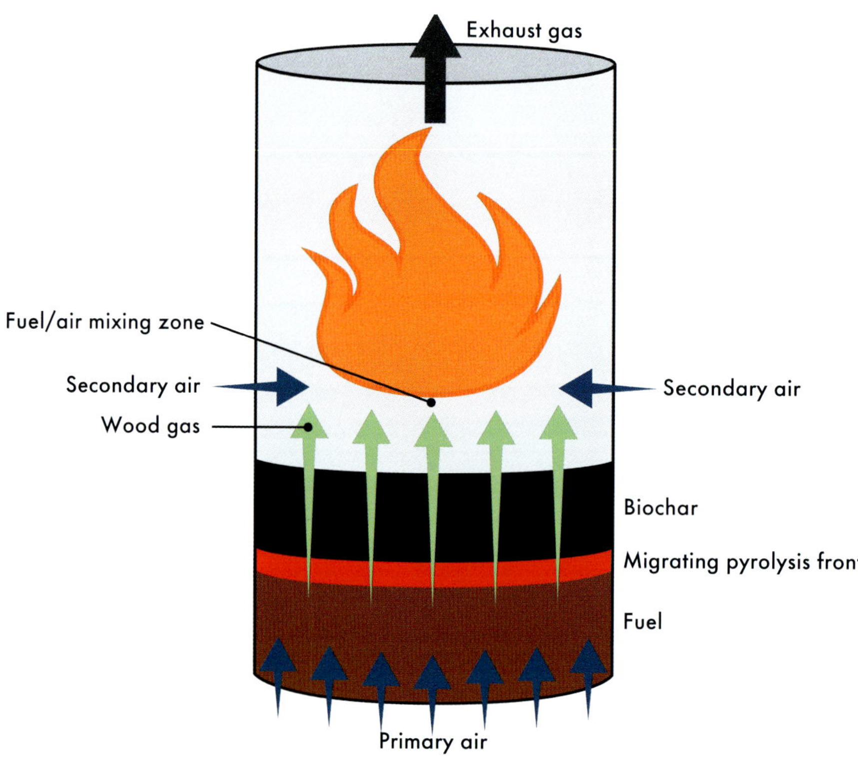

Figure 4.4. How a TLUD works. 1. Primary air flows up through the fuel, releasing wood gas and forming char. 2. Wood gas rises up through the char to the fuel/air mixing zone. 3. Wood gas comes into contact with secondary air, where it mixes and burns in a flame. 4. Exhaust gas exits out the top.

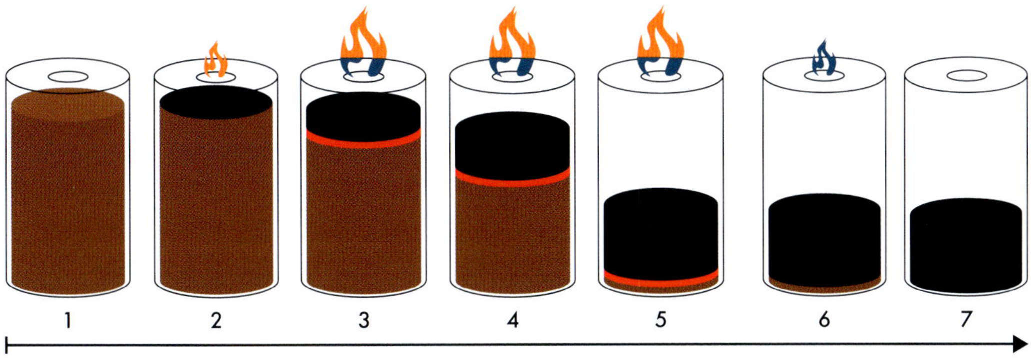

Figure 4.5. A time series that illustrates how the pyrolysis front (red) migrates down through the fuel column, leaving behind a charcoal residue. Adapted from image courtesy of GTZ

As the pyrolysis front migrates downward, it keeps igniting the new layers below it, forming more char and releasing more wood gas. Meanwhile, the wood gas rises up through the already formed char where it meets and mixes with the secondary air, burning in a flame. The exhaust from the burning flame, mostly CO_2, rises and exits out the top. Figure 4.4 shows the flows of primary air, secondary air, and exhaust gas.

Figure 4.5 is a time series that illustrates how the pyrolysis front migrates down through the fuel until all the fuel has been gasified and nothing is left but the char residue. This illustration is from a very useful design guide by international development expert Christa Roth, titled *Micro-Gasification Cooking with Gas from Dry Biomass: An Introduction to Concepts and Applications of Wood-Gas Burning Technologies for Cooking.*[2] You can find this guide and many other useful TLUD and biochar materials online at Dr. Paul Anderson's website: DrTLUD.com.

Different TLUD designs are more or less successful at producing both biochar and heat for cooking. An ideal TLUD will optimize both functions. Closeness to the ideal depends on maintaining a balance of forces, the three Ts of combustion: time, temperature, and turbulence.

Complete and efficient combustion can be enhanced by:

1. Lengthening the *time* that air and wood gas remain in the combustion zone (residence time)
2. Raising the *temperature* of the reactants (air and fuel)
3. Increasing *turbulence* to promote thorough mixing of air and fuel

Residence time in the combustion zone in a natural-draft TLUD is partly determined by the velocity of the primary airflow. This will be determined by a balance between buoyancy and friction. Even without a temperature difference, buoyancy arises from the pressure difference between air at the intake of a furnace or stove and air at the outlet, which is physically higher than the intake and where the atmospheric air pressure is slightly lower. This is called the stack effect. The taller the stack, the greater the pressure differential and the greater the airflow up the stack. If the air is heated, it will expand and increase the pressure differential even more. This is why a woodstove has a tall chimney. The tall chimney creates more draft to feed the fire in the stove with air pulled in from below.

Here are some of the ways you can optimize the 3 Ts in your stove design:

Time. A combustion chamber can be designed to be just the right size and shape to balance draft with friction resistance so the air and fuel mix in a "bubble" of hot gas and flame underneath a pot for cooking, increasing the residence time.

Temperature. A way to further improve combustion is to increase the temperature of the reactants by designing a combustion chamber that is better insulated, usually by using ceramics.

Turbulence. One way to increase both time and turbulence is to add a little swirl to the combustion zone. The swirl expends the kinetic energy of rising gases within the combustion chamber, while also enhancing turbulence and fuel/air mixing.

Three Stove Designs

I had a lot of fun playing with what the stove designer community calls "tin-canium," the ubiquitous tin can material that is easy to obtain from your own trash. If you want to use the big number-10 cans, just ask a restaurant. A quick cruise through town stopping at half a dozen Mexican restaurants yielded more than thirty large, empty refried bean cans for our kids' camp one year. Using simple hand tools, you can try out many different designs to see what you like best. After many happy hours playing around, I came up with a few designs that optimize the three Ts and are functional to cook on. The three basic stove designs shared here are: the Bubble Stove, the Whirly Girl, and the Whirly Pinay.

THE BUBBLE STOVE

I developed the Bubble Stove as a fun biochar project for kids at a summer camp, but it is also a functional stove and very simple to make. All you need is two tin cans, four sheet-metal screws, a large nail, and a hammer or similar tool (kids at the camp just used rocks).

Tools and Supplies

one #10 tin can—ask a restaurant to save them for you
one 28-ounce (800 ml) tin can—canned tomatoes and peaches come in this size
hammer

one large nail—a 16d or "16-penny nail" works best
four sheet-metal screws
can punch—the kind you use to open a juice can
screwdriver (that works with your sheet-metal screws)
pliers
permanent marker
ruler or straight edge
wire coat hanger—optional, for making a handle to lift out the inner can when it is hot

Step 1. Prepare Inner Can

Start with the smaller can. Remove any paper labels. With your permanent marker, indicate where all the holes will go, bottom and top. Punch nail holes on the bottom, ensuring and maybe marking four evenly spaced holes near the rim for the legs. Then take the nail and punch many small holes all over the bottom of the can for the primary air, as shown in figure 4.6.C. Install the four sheet-metal screws as legs using a screwdriver. Make sure they are even, so that the can will be level. Finally, use the can punch to make eight large evenly

Figure 4.6. Step one of Bubble Stove construction: **A.** Find the right two cans. **B.** Using the nail, punch holes across the bottom of the small can. **C.** Install screws in four of the holes to make legs. **D.** Punch eight triangular holes around the top rim of the small can.

spaced holes around the top for secondary air. With some pliers, wrap the sharp tips of tin around the rim so they don't cut anyone.

Step 2. Prepare Outer Can and Assemble

Remove the label on the large #10 can. All the holes in this can are made using the can punch. There are no nail holes. Using the can punch, make sixteen evenly spaced holes around the top (open) end of the can and eight holes around the bottom end of the can. Using the pliers, wrap the sharp tips of tin around the rim holes. Then place the small can, loaded with fuel, inside the big can. Try to center it. That is the entire stove.

Step 3. Fuel and Light the Stove

The Bubble Stove is designed to work best with commercial fuel pellets. You can find these at many hardware and home improvement stores. They are used to fuel pellet stoves and furnaces for home heating. But you can also try some other fuels that meet these basic requirements:

- Small, evenly sized pieces
- Dry
- Not packed too tightly (ensuring that air can flow)

Figure 4.7. Step two of Bubble Stove construction: Using the can punch, make sixteen evenly spaced holes around the top (open) rim of the large can and eight evenly spaced holes around the bottom (*left*). Then place the small can, loaded with fuel, inside the large can (*right*).

Fuels that will not work well are sawdust and small wood chips—they pack too tightly and do not allow air to flow up from the bottom of the can. Some fuels you might try include:

- Crushed pine cones
- Small twigs
- Nut shells
- Fruit pits
- Chopped-up corn stalks or cobs

Some fuels are much harder to light than others. Pellets are actually fairly difficult to light because they are so dense. A less dense fuel like straw will be much easier to light, but it will not burn nearly as long as wood pellets. Here are some easy aids to help light the pellets:

- Cotton balls soaked in alcohol
- Candle wax dripped on top of the pellets
- A few squirts of fire-starting gel
- A stack of small, dry twigs on top of wood pellets

Step 4. Cook Something

You can cook anything that uses a "medium-high" cooking flame. You can boil water for tea or coffee. You can cook a pot of rice or soup. You can fry eggs or meat in a frying pan. The stove will boil 2 quarts (2 L) of water in about 20 minutes. If you are using wood pellets, cooking time will last about 45 minutes.

Step 5. Quench and Save the Biochar

As the stove gets to the end of the fuel, you will see the flame get weaker and weaker. Finally, it will go out and you will see a tiny puff of smoke. If you just leave it, the char will slowly burn with no visible flame and the top of the fuel will start to turn white with ash. To save the char, use leather gloves and pliers, or a handle made of a bent coat hanger, to lift the inner can out of the stove and gently drop it in a bucket of water or douse it with a hose. After the char cools, strain it and measure it. Once it is dry you can put it on a scale and weigh it. See how it compares by weight and volume with the wood pellet fuel. Your char may be only a little less than half the volume of the pellets, but it will weigh a lot less.

Figure 4.10 on page 77 shows how the Bubble Stove works. Primary air comes through the eight holes punched around the bottom of the large can. Part of the air flows up through the nail holes in the bottom of the inner can, where it feeds the migrating pyrolysis front and then sweeps the wood gas up into the burn chamber. The rest of the air flows through the annular gap between the two kilns where it is heated before it enters the eight punch holes around the rim of the small can. This preheated secondary air then mixes with the wood gas and burns in a flame. The flame exhaust gas then exits the sixteen punch holes around the rim of the large can, rising up around the cooking pot.

Figure 4.8. **A.** Cotton balls soaked with rubbing alcohol are placed on top of the pellets and lit (**B**). **C.** Once the pellets are burning, we place a tea kettle on the stove. **D.** When you lift the tea kettle off the stove you can see how the jets of secondary air coming though the top holes in the inner can form clean-burning flamelets.

Flame shaping in the Bubble Stove is not extremely precise, but under most conditions, the stove will burn with very little smoke. Keeping the flame right under the pot and feeding the flame with preheated air are the keys to clean and efficient cooking with the Bubble Stove.

Use these basic concepts to design your own Bubble Stove using the materials and fuels that are available to you. For instance, you could make a larger stove using bigger cans like metal drums or metal trash cans. With a larger stove you can use larger feedstocks like wood chunks or sticks. If you need to make larger holes, use a drill or a rotary cutting tool. Remember that you need to match the stove geometry to the geometry of the fuel.

Figure 4.9. The flame has gone out and the char is ready to be quenched (*left*). The charred pellets shrink by half from the starting volume (*right*).

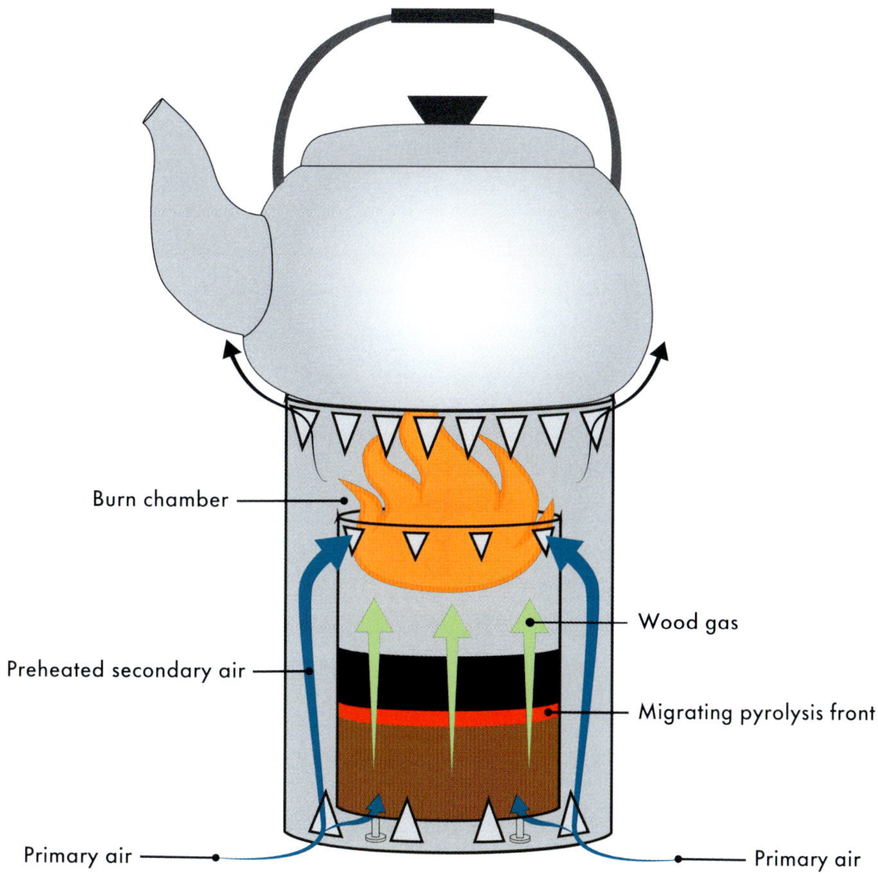

Figure 4.10. Air and gas flows in the Bubble Stove.

THE WHIRLY GIRL

The Whirly Girl stove will run for over an hour on a full charge of wood pellets, so you can get a lot of cooking done. If you use it as a space heater by adding a stack and a stove pipe, the extra draft will cause the pellets to burn more quickly and intensely. You will likely end up with less char and have to refuel more often, but you will get more heat. The following are the steps to make the Whirly Girl.

Start with two large cans like the #10. The bottom can (fuel can) is made similar to the Bubble Stove inner can: use a nail and hammer to punch primary air holes in the bottom and install screws for feet. But then punch four additional evenly spaced holes around the top rim and attach some large washers. These are for securing the top can, which is the burn chamber.

Figure 4.11. Steps for making the Whirly Girl stove: **A.** Prepare the bottom can with bottom air holes, feet, and washers to secure the top can. **B.** Mark where you will cut slits to make louvers and the hole at the top, then fill with water and freeze. **C.** Use a nail to punch a hole at the beginning and end of each cut. **D.** Use a rotary tool to cut slits and tabs for the top hole.

The burn chamber is made with precision cuts using a rotary tool with a cutting wheel. To make the cuts precisely, it helps to start by making a nail hole at both ends of each cut line. Because it is hard to make holes in the sides of a can without denting or crushing it, first mark where all the cuts will go and then fill the can with water and put it in the freezer. Then you can hammer nails into the can against the solid ice without denting the can.

Once the nail holes are in place, slice two rows of slits in the sides of the can with the rotary tool, and then use it to cut out a hole in the top. Dump the ice out of the can. If you want to make a patio heater, bend the top hole slices up so you can slip a stove pipe over them. If you want to make a cooking stove, bend them down and back out of the way, or just cut them off.

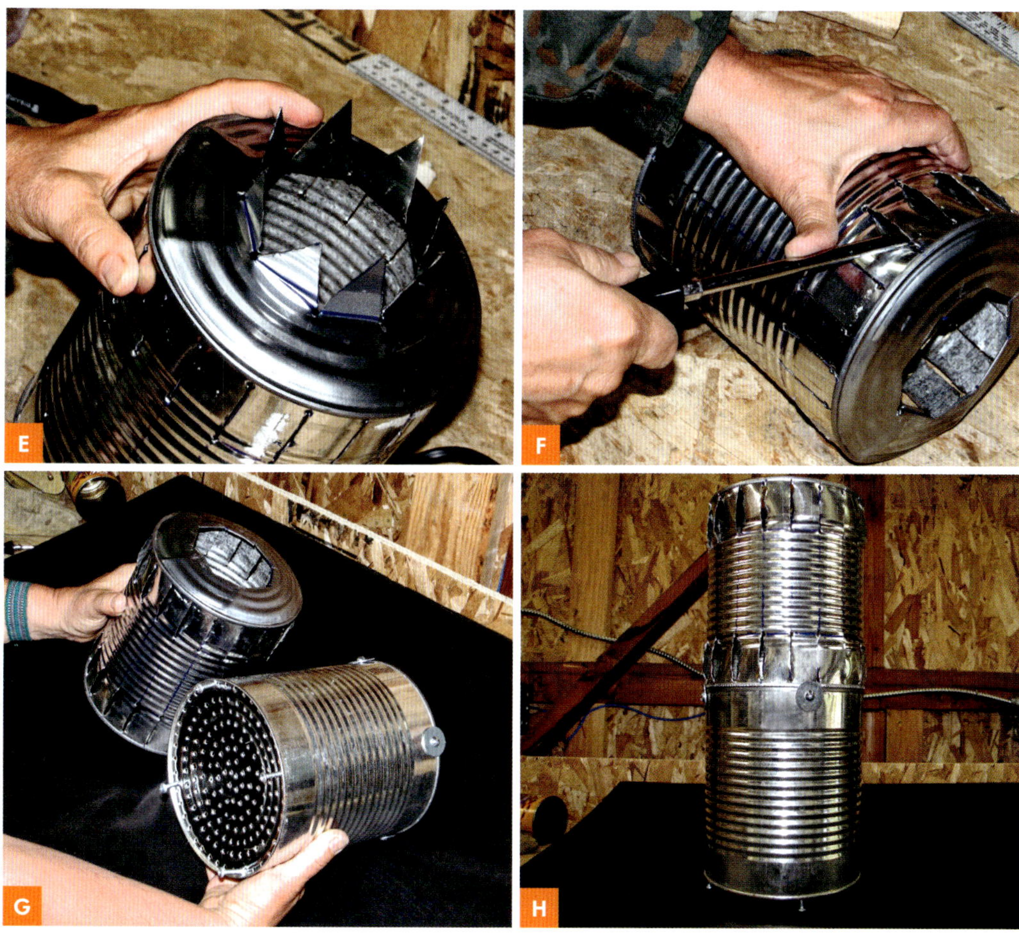

E. Bend the tabs back to open up the top hole. **F.** Use a screw driver to spread the slits open to form louvers. **G.** The two completed cans are ready to combine into the finished stove. Washers on the bottom can will keep the top can from sliding off. **H.** Finished stove.

Figure 4.12. Using a two-burner Whirly Girl to cook burgers and beans, indoors under a hood.

Using a screwdriver, spread open up all of the slits around the top can to make louvers, like you see in an air vent. These will direct secondary air into the burn chamber in a swirl pattern. You will be able to see the swirling flames in the burn chamber. Figure 4.11 illustrates these steps.

To use the stove, you need to set up a grate above the stove as in Figure 4.12. Load up the bottom can with wood pellets and light, then set the burn chamber on top of the fuel can. Primary air comes up through the bottom of the fuel can and secondary air all comes in through the louver slits. We cannot preheat the secondary air in this design, but we do gain the advantage of the swirling combustion air turbulence for better fuel/air mixing. Adding a stove pipe increases the velocity of the airflow and the stove will put out a lot of heat for your patio or even a greenhouse in the winter.

THE WHIRLY PINAY

I met Rebecca Vermeer at a biochar school we held in California with the Sonoma Biochar Initiative in 2014. Rebecca is a Pinay—a Filipina woman—who had moved to Canada but wanted to help people in her home country become more self-reliant. As in other parts of the world where cooking fuels are scarce and expensive, improved cookstoves can make a big difference particularly in the lives of the women, who do most of the cooking. Rebecca was also interested in helping them make biochar in cookstoves to enhance their ability to produce food from their gardens and farms.

Rebecca and I had a fruitful correspondence about the Whirly Girl stove and she wanted to adapt it for construction with the plentiful local clay. The design that she came up with was simple, elegant, and very effective. The ceramic material is insulative so both the fuel chamber and the burn chamber run hotter and cleaner. With the added swirl in the burner, it's a great stove. However, the ceramic is not the toughest material, and with repeated handling, the parts can crack. An improved ceramic could make this stove a real winner.

TLUD stoves come in many shapes and sizes and there are designs that use fan assist for precise control of primary and secondary air. When used with

Figure 4.13. The Whirly Pinay cooks a meal of meat and vegetables with plenty of coals left over to quench into biochar.

consistent fuels like pellets made from crop or wood waste, they can rival a propane stove for clean combustion, and with proper ventilation are safe to use indoors. Where a local industry can be established to provide stoves and pellet fuel to cooks, whole communities can become self-reliant in clean cooking without fossil fuels or electricity, while producing biochar for household use. Fuel resources like oil, gas, coal, and electricity are not evenly distributed globally. Most parts of the world, though, have enough waste biomass available to feed a pellet supply industry for clean cookstoves.

The Drum TLUD and the Big Whirly

TLUDs can provide heat energy for cooking, but sometimes you just want to make biochar. In that case, you can make a large TLUD out of a barrel. The Big Whirly is a variation on the 55-gallon-drum TLUD that many people have used for their first biochar production for gardens and small farms. You will find lots of examples of the drum TLUD across the internet.

The Big Whirly differs in two ways from the more common designs. First, we cut slits in the burn chamber to get some swirling secondary air, but perhaps more importantly, we use an extra-large burn chamber, without a stack. Recent research on TLUD combustion dynamics confirms that designs with larger burn chambers work better.[3] The burn chamber is made of an entire second drum, placed on top of the fuel chamber. The burn chamber drum sits directly on the fuel chamber drum, with some metal tabs to hold it in place. I added some handles to the burn chamber to make it easier to lift on and off. The exhaust gas exits the burn chamber through three rectangular holes cut around the top of the drum with a cutting torch. The top of the burn chamber drum is closed, like a chimney cap.

The fuel chamber is constructed just like the one for the small Whirly Girl made out of a tin can, except we use a drill to make holes in the bottom of the drum. We place a few bricks or rocks under the drum to serve as legs or spacers to allow primary air to enter from the bottom holes. If it seems like there is

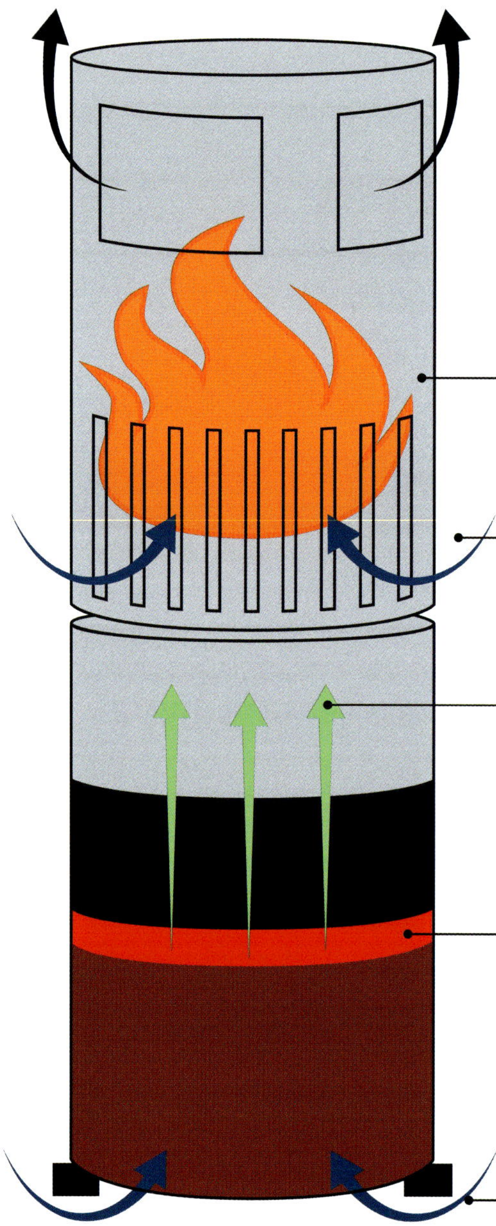

Figure 4.14. Schematic showing air and gas flows in the Big Whirly barrel TLUD.

too much primary air entering the fuel, we use smaller rocks. Figure 4.14 is a diagram showing the airflows in the Big Whirly.

If the fuel chamber is packed full of fuel, flames will fill the burn chamber at the beginning of the burn and tend to shoot out the top as shown in the first TLUD design in figure 4.15. As the fuel level burns and drops lower in the kiln, the flames will die down. Better, more complete combustion of the wood

Figure 4.15. The first TLUD (*top*) has a nice swirling combustor, but it could be larger. Once the fuel burns down a bit, the flames will stay in the burner. The Big Whirly TLUD (*bottom*) has a larger burn chamber that keeps flames contained throughout the burn.

gas can be achieved by providing a bigger burn chamber as we have in the Big Whirly design on the bottom.

If you wish to design your own barrel TLUD, keep these points in mind:

- Design the TLUD for the fuel feedstocks that you have, keeping in mind the amount of air spaces between fuel pieces and the feedstock packing density.
- Make sure your fuel is as dry as possible. Wet feedstocks will not work.
- Try to make the fuel as uniform as possible for the most efficient burn.
- Adjust the air flow by controlling the primary air at the bottom. Place the drum on bricks to allow more airflow. Densely packed fuels will need more primary air. Light fuels like hemp stalks will need very little.
- Avoid any constriction in the design that could cause back pressure and put the flame out.
- To avoid snuffing the flame while getting started, make sure that the fuel in the fuel drum is burning strongly before you place the burner on top of the fuel drum.
- If you lose the flame, a TLUD becomes a smoker and you will need to quickly relight it using a strong flame from a propane torch. A match will not do the job.

Making Biochar in Existing Furnaces

Another way to make biochar in a flaming combustion process is to make it in a furnace. A furnace is simply an insulated burn chamber, with or without an integrated heat exchanger, that accepts biomass loaded manually or through automated systems. Woodstoves are common examples. Industrial wood-fired furnaces are often called "boilers," and there are many different designs and sizes. Smaller furnaces mostly have fixed beds, while large furnaces often have fluidized beds to allow clean combustion of large amounts of fuel.

Here's a little secret about the biochar industry in the United States today: The biggest commercial biochar producers are retrofitted biomass power plants. Biomass power plants use ground up wood and operate much like a coal-fired power plant: Fine-ground fuel is blown into a furnace with boiler tubes. The fuel burns and heats the water to make steam that drives a steam turbine to produce electric power. The exhaust gas continues up through a stack, but it first goes through a cyclone filter to drop out large particulates and then a series of other pollution controls to filter out fine particulates. With modern pollution controls, these wood-fired power plants are very clean. But the interesting thing about them is what happens to the large particulates. Because the transit time in the furnace is unavoidably fast, these particulates are not just made of mineral ash. They are also partly composed of carbon that charred but did not burn completely.

In the Pacific Northwest, power plants and furnaces at sawmills that are used to dry lumber have been producing mountains of this "high-carbon ash" from their cyclone filters. Instead of paying to landfill it, they have quietly offered it to farmers. When I started reaching out to farmers in Oregon about the benefits of biochar, I soon ran into some who had used biochar for years already in the form of this ash. We tested a few samples of it and found that it was about 40 percent carbon.

In Canyonville, Oregon, Michaels Ranch produces grass-fed beef on pasture soils that are low pH (between 5.2 and 6.0) and somewhat low in potassium. Several years ago, Troy Michaels began using high-carbon boiler ash as a liming agent and as a supply of potassium. Boiler ash is an economical alternative to lime, which is very expensive in Oregon. Lime spreading costs $100 an acre or more for one ton of lime. While the high-carbon ash has only about one-seventh of the liming value of lime, it is free and the sawmills will deliver it to the farm. Michaels noticed that the fields where he had applied boiler ash greened up faster in the spring, so he was very interested in learning more about biochar.

Most users of biochar do not want to apply that much ash, so some of these boiler operators have figured out how to filter out the ash component to produce a biochar product that is close to 85 percent carbon.

One biomass powerplant that has jumped into the biochar business is Biomass One, a 30-megawatt biomass power plant in White City, Oregon. They can produce 3,500 to 4,000 tons of biochar a year with no impact on the power output of the plant. It requires a very slight increase in the amount of biomass

Figure 4.16. Troy Michaels and a visitor with a giant pile of high-carbon ash that will be applied to his pasture.

fuel. The largest share of the company's revenue comes from electricity sales, followed by tipping fees from their green waste drop-off service. Biochar and a soil-blending yard make up the last piece of the revenue stream.

Karl Strahl, the vice president in charge of biochar sales, said that even though they make money selling electricity, it makes more sense to think of Biomass One as a wood-waste recycling plant that recovers value from waste in the form of electricity and biochar. Biomass One is a great example of how we can squeeze every last bit of value out of what we often consider waste. In Oregon, where woody debris is abundant, and since transportation costs will never support the export of wood chips, being able to fully use this resource locally is an important contribution to our self-sufficiency.

Smaller biomass furnaces can also make biochar while producing energy. The term for this technology is "combined heat and biochar," or CHAB. While a plant that produces electricity would cost many millions of dollars to site and build, smaller units that provide heat can be had for much less. CHAB technologies can include retort systems, but gasifiers and furnaces are more common. One example of a simple, robust, but powerful CHAB furnace is a unit from BioMass Energy Techniques (BET), a Mennonite-owned company.

The basic BET furnace design exists in dozens of installations, mostly at sawmills, to provide heat for lumber-drying kilns. In its CHAB configuration, it uses an inclined grate with a char removal auger below the grate. The long, highly insulated burn chamber and limited primary combustion air produce gasification conditions that can promote water gas reactions to liberate burnable hydrogen from water. This allows use of feedstocks with higher moisture content. In this furnace, green wood chips provide more heat energy than fully dried wood chips, and optimum moisture content is a minimum of 25 percent.

The open grate is designed so that as the wood chips char and shrink, they will fall through the grate to a char removal auger before they burn completely to ash. An induction fan controls airflow. The furnace is designed for very low airflow, which means low transport of particulates and clean flue gas. The furnace can be supplied with an integrated boiler or a flue gas-to-air heat exchanger.

I was able to visit a sawmill in Oregon that has one of the BET furnaces. This furnace does not make biochar, but it was interesting to see how simple and robust it is. The furnace has very few moving parts, which is important for reliability and reduced maintenance needs. Adding a char removal auger would add more moving parts, but the company has a track record of manufacturing reliable equipment that is in service for decades.

The BET 49-S model that makes biochar supplies 3.5 million BTU (British thermal units) of heat per hour while producing 214 cubic yards (164 m^3) of biochar per month of operation. The heat in one BTU equals roughly the heat supplied by one wooden kitchen match. To visualize this, think of how many million

Figure 4.17. The Biomass Energy Techniques furnace uses wood chips fired on an inclined grate. Char and ash tumble to the bottom, and the biochar version of the furnace uses an auger to remove the char before it burns completely to ash.

kitchen matches it would take to heat a giant warehouse in the middle of winter in Nebraska.

To see a comparison of different CHAB technologies, you can read a report that I produced for the Nebraska Forest Service and the US Biochar Initiative (USBI), with the help of Tom Miles, USBI executive director: "Combined Heat and Biochar Technology Assessment for a Composting Operation."[4] This report looked at technologies available from five different vendors that could provide heat and biochar for Big Red Worms, a worm farm and composting facility in Lincoln, Nebraska. The worms are in a 4,000-square-foot (372 m^2) warehouse that is heated to protect them from harsh Nebraska winters. The peak heating requirement for the minimally insulated warehouse is in the range of 500,000 to 750,000 BTU per hour.

A CHAB system that provides heat for the worms and biochar to improve the composting operation could add a lot of value. Current heating bills to warm the worms run about $1,500 a month in the winter. Wood chips are free and abundant, while the price of propane keeps going up. Add in the biochar and its value in producing more and better compost, and a CHAB unit could easily pay for itself. And there is no issue about providing enough heat while making biochar. The smallest CHAB unit I looked at produced one million BTU an hour. The largest could provide eight million BTU.

Even if you don't need a couple of million BTU of heat for your farm or homestead, there are still ways to combine biochar and heat. If you already heat your home or shop with a wood stove, there are two ways to make biochar and heat. The first way is to put wood chips or chunks into a small retort and load it into the stove. You can use a clean paint can with lid. Punch a few holes in the lid for the gas to escape and load it into your wood stove like you are putting another log on the fire. After a while you will see smoke coming out of the holes, and if your stove has a window in the door, you can watch the smoke turn into small jets of flame. Compared to what's on TV, that's pretty good entertainment!

The other way to make biochar in your wood stove is to shut the stove down at night when there is a good charge of half-burned wood. If your stove is airtight, the wood will pyrolyze without burning to ash. In the morning you can scoop out the chunks of char and save them in a bucket. Make sure they are completely out and cold. I have one friend who does this all winter and ends up with plenty of biochar for his garden.

I think it is fair to assume that throughout human history, people have saved pieces of char from their cooking fires. Mothers probably knew that a bit of crushed char could relieve a child's tummy ache, and that char would help a wound to heal. Char would deodorize a slop bucket on its way to the garden. Middens surrounding ancient villages and towns are full of charcoal that is still there today for archeological digs to find. Now we know that char is not just a convenient substance for carbon dating an ancient human habitation. It was also likely something that was deliberately used for multiple benefits to human health, sanitation, and food security. And its use was so common, that no one bothered to record it in writing.

CHAPTER FIVE

Biochar Properties

Now that you have some biochar, what are you going to do with it? Before you shovel loads of biochar into your garden beds, you need to know a bit more about it.

Biochar is more than just carbon. Biochar made from wood can contain as much as 90 percent carbon, 8 percent other organics, and 2 percent ash, while biochar made from corn stover, grass, or seed hulls may contain only 70 percent carbon, 10 percent other organics, and 20 percent ash. A lot depends on how the biochar is made. Higher temperatures and too much air will produce more ash. While we certainly want lots of carbon in our biochar, ash is useful as well. Ash is just the mineral components of the biomass—calcium, potassium, phosphorus, silica, iron, and many trace metals. Most of these minerals are considered to be fertilizer or plant food. In general, if the plant took these minerals up from the soil, then they were wanted or needed by the plant. If we can return them to the soil, we are doing the plants a favor.

The other organics can be many different organic molecules that contain carbon, hydrogen, oxygen, nitrogen, and other volatile elements that did not get completely cooked out of the biochar. These are called volatile matter because they can become a gas when heated. Technically, even the biochar is a form of volatile matter that can also burn and become carbon dioxide or carbon monoxide; however, we have gone to a lot of trouble to convert it to a more stable form of carbon, so we call that fraction of the biochar "stable carbon," "fixed carbon," or "recalcitrant carbon." As far as we can tell, that stable fraction can last for thousands of years when buried in soil.

Most species of wood produce biochar with fairly similar qualities. Biochar made from straw, stover, rice hulls, manure, or other materials will have different qualities. Generally, straw biochar has more ash and less carbon, rice hull biochar is high in silica, and manure biochar has more nitrogen and more ash than wood biochar.

Table 5.1. Proximate Analysis and Stability Index for Two Different Biochar Feedstocks Made at Five Different Temperatures

Feedstock	Production Temperature (°C)	Fixed Carbon (%)	Ash (%)	Volatile Matter (%)	H:C
Ponderosa pine wood	300	28.2	1.5	70.3	1.42
	400	62.2	1.4	36.4	0.8
	500	72.7	2.1	25.2	0.52
	600	85.2	3.7	11.1	0.4
	700	92	1.7	6.3	0.21
Tall fescue grass	300	36.2	9.4	54.4	1.34
	400	56.9	16.3	26.8	0.73
	500	64.3	15.4	20.3	0.48
	600	67.6	18.9	13.5	0.33
	700	71.6	19.3	9.1	0.2

Adapted from Marco Keiluweit et al., "Dynamic Molecular Structure of Plant Biomass-Derived Black Carbon (Biochar)," *Environmental Science & Technology* 44, no. 4 (2010): 1247–53.

It's important to know that biochar made from the same feedstock has different qualities when made at higher or lower temperatures. Let's look at some data from two different feedstocks, a woody feedstock (ponderosa pine) and a grass (tall fescue), that were made into biochar in the laboratory furnace at five different temperatures.[1] Table 5.1 gives the percentages of fixed carbon, ash, and the remaining volatile matter that did not get converted to char. These percentages add up to 100 percent. This analysis is called the "proximate" analysis. The proximate analysis terminology comes from the coal industry, where *proximate* analysis gives the simple fractions of water, ash, fixed carbon, and volatile matter, while *ultimate* analysis gives the proportions of the main elements: carbon, hydrogen, oxygen, and nitrogen.

The table also includes a stability index called the H:C ratio, which is the ratio of hydrogen to organic carbon in the material. The bigger the number, the less stable the biochar. The cutoff value is 0.7. A material with H:C greater than 0.7 is not considered to be biochar. By that standard, the pine and grass examples in table 5.1 that were processed at temperatures at or below 400°C would not qualify as biochar.

It is useful to know what kind of biochar you have, because it will help you know how to use it best. You can get a thorough analysis of biochar by sending it to a lab, and we will look at how to interpret biochar lab test results. You can also do a basic fractional analysis at home using a toaster oven and a propane-fired camp stove. The procedure for doing this was developed by activated carbon chemist Hugh McLaughlin, and is explained in detail in *The Biochar Revolution*, a helpful handbook of biochar tips and tricks edited by Paul Taylor.[2] This book is so useful that I recommend that every serious biochar experimenter own a copy of it.

The basic proximate analysis procedure is to first dry a sample of biochar in the toaster oven to remove the water. Weigh the biochar before and after drying to find the weight of the water. Then take the dry sample and cook it at a high temperature in a tuna can over the propane stove until it is burned to ash. Again, weigh it before and after and subtract the ash weight to get the carbon weight. There is another procedure that is a little trickier that involves cooking the dried biochar at a lower temperature to volatilize the organic molecules remaining in the biochar. You then weigh the remaining devolatilized char and subtract that from the original dry biochar weight and the ash weight, to get the weight of the volatile matter. Another term for the volatile matter is "mobile matter" because this is the component of the biochar that will be easily consumed by microbes once the char is added to soil or compost.

There is a lot more to know about all three of the basic biochar fractions. The volatile matter could include chemicals that enhance or inhibit plant growth. The ash could include nutrients that you need, or it could include heavy metals that you don't want. The char could be very porous with lots of surface area, or the pores could be clogged leaving little available surface area.

Biochar Standards and Testing

The International Biochar Initiative has developed a biochar testing and certification program with three panels of testing: Basic Utility Properties, Toxicant Reporting, and Advanced Analysis and Soil Enhancement Properties. In the United States, Control Laboratories in Watsonville, California, is able to do all of the required tests except the very expensive toxicant tests for dioxins, furans, PCBs (polychlorinated biphenyls), and PAHs (polyaromatic hydrocarbons). Fortunately, these toxicants are not usually found in biochar made from clean woody debris. To learn more about biochar characteristics, let's take a look at some lab reports.

Figure 5.1 (sample A) is a lab report on biochar characteristics from a Ring of Fire kiln used with mixed woody debris in Montana. Figure 5.2 (sample B) is a report from the same lab on Ring of Fire biochar made from mixed woody debris in California. Let's go through each test to understand what these reports are telling us, why we would want to know this information, and how the two reports differ.

Moisture. The percentage of the sample that is water. The sample is dried to remove the water before further analysis. One sample is very wet and the

International BioChar Initiative (IBI) Laboratory Tests for Certification Program

	Dry Basis Unless Stated:	Range	Units	Method
Moisture (time of analysis)	60.9		% wet wt.	ASTM D1762-84 (105c)
Bulk Density	9.2		lb/cu ft	
Organic Carbon	84.9		% of total dry mass	Dry Combust-ASTM D 4373
Hydrogen/Carbon (H:C)	0.27	0.7 Max	Molar Ratio	H dry combustion/C(above)
Total Ash	8.0		% of total dry mass	ASTM D-1762-84
Total Nitrogen	0.57		% of total dry mass	Dry Combustion
pH value	8.32		units	4.11USCC:dil. Rajkovich
Electrical Conductivity (EC20 w/w)	0.083		dS/m	4.10USCC:dil. Rajkovich
Liming (neut. Value as-CaCO3)	7.5		%CaCO3	AOAC 955.01
Carbonates (as-CaCO3)	2.9		%CaCO3	ASTM D 4373
Butane Act.	6.4		g/100g dry	ASTM D 5742-95
Surface Area Correlation	335		m2/g dry	G

All units mg/kg dry unless stated:		Results	Range of Max. Levels	Reporting Limit (ppm)	Method
Arsenic	(As)	**ND**	13 to 100	0.49	J
Cadmium	(Cd)	**ND**	1.4 to 39	0.20	J
Chromium	(Cr)	3.9	93 to 1200	0.49	J
Cobalt	(Co)	**ND**	34 to 100	0.49	J
Copper	(Cu)	10.0	143 to 6000	0.49	J
Lead	(Pb)	0.81	121 to 300	0.20	J
Molybdenum	(Mo)	7.3	5 to 75	0.49	J
Mercury	(Hg)	**ND**	1 to 17	0.001	EPA 7471
Nickel	(Ni)	2.0	47 to 420	0.49	J
Selenium	(Se)	**ND**	2 to 200	0.98	J
Zinc	(Zn)	225	416 to 7400	0.98	J
Boron	(B)	15.6	Declaration	4.9	TMECC
Chlorine	(Cl)	98.6	Declaration	20.0	TMECC
Sodium	(Na)	1164	Declaration	491	E
Iron	(Fe)	368	Declaration	24.5	E
Manganese	(Mn)	289	Declaration	0.49	J

Particle Size Distribution	Results	Units	Method
< 0.5mm	1.7	percent	F
0.5-1mm	0.8	percent	F
1-2mm	0.8	percent	F
2-4mm	2.8	percent	F
4-8mm	9.0	percent	F
8-16mm	30.3	percent	F
16-25mm	54.7	percent	F
25-50mm	0.0	percent	F
>50mm	0.0	percent	F
Basic Soil Enhancement Properties			
Total (K)	7372	mg/kg	E
Total (P)	2030	mg/kg	E
Ammonia (NH4-N)	22.3	mg/kg	A
Nitrate (NO3-N)	1.9	mg/kg	A
Organic (Org-N)	5679	mg/kg	Calc.
Volatile Matter	14.3	percent dw	D

* "ND" stands for "not detected" which means the result is below the reporting limit.

Method A Rayment & Higginson
D ASTM D1762-84
E EPA3050B/EPA 6010
F ASTM D 2862 Granular
G Butane Activity Surface Area Correlation Based on McLaughlin, Shields, Jagiello, & Thiele's 2012 paper: Analytical Options for Biochar Adsorption and Surface Area
J EPA3050B/EPA 6020

Figure 5.1. Biochar test results for sample A.

other is dry. Both were quenched with water, so most likely, the char from sample B sat in a pile in the sun and dried before it was sampled.

Bulk Density and Particle Size Distribution. The density of the provided sample. This is a tricky one because of particle size. Completely carbonized biochar made at a high temperature is brittle and breaks easily, but it often starts out as a sizeable chunk. You can easily pull a sample that includes several fist-sized chunks, surrounded by small granular-sized pieces, along with biochar dust. Bulk density is measured by weighing a known volume of material to get the pounds per cubic foot. But that will change if the size of the particles changes. A bucket of rice will have the same bulk density every time you measure it. A bucket of biochar can change. Try it yourself by weighing a bucket of your biochar. Then smash it up as fine as you can with a shovel or a brick and fill the bucket again. You might have reduced the volume by half and so doubled the bulk density.

International BioChar Initiative (IBI) Laboratory Tests for Certification Program

	Dry Basis Unless Stated:	Range	Units	Method
Moisture (time of analysis)	5.9		% wet wt.	ASTM D1762-84 (105c)
Bulk Density	13.4		lb/cu ft	
Organic Carbon	86.8		% of total dry mass	Dry Combust-ASTM D 4373
Hydrogen/Carbon (H:C)	0.37	0.7 Max	Molar Ratio	H dry combustion/C(above)
Total Ash	4.9		% of total dry mass	ASTM D-1762-84
Total Nitrogen	0.59		% of total dry mass	Dry Combustion
pH value	8.40		units	4.11USCC:dil. Rajkovich
Electrical Conductivity (EC20 w/w)	0.081		dS/m	4.10USCC:dil. Rajkovich
Liming (neut. Value as-CaCO3)	10.7		%CaCO3	AOAC 955.01
Carbonates (as-CaCO3)	3.1		%CaCO3	ASTM D 4373
Butane Act.	6.1		g/100g dry	ASTM D 5742-95
Surface Area Correlation	327		m2/g dry	G

All units mg/kg dry unless stated:		Results	Range of Max. Levels	Reporting Limit (ppm)	Method
Arsenic	(As)	**ND**	13 to 100	0.28	J
Cadmium	(Cd)	**ND**	1.4 to 39	0.11	J
Chromium	(Cr)	1.5	93 to 1200	0.28	J
Cobalt	(Co)	0.67	34 to 100	0.28	J
Copper	(Cu)	2.8	143 to 6000	0.28	J
Lead	(Pb)	0.41	121 to 300	0.11	J
Molybdenum	(Mo)	**ND**	5 to 75	0.28	J
Mercury	(Hg)	**ND**	1 to 17	0.001	EPA 7471
Nickel	(Ni)	1.9	47 to 420	0.28	J
Selenium	(Se)	**ND**	2 to 200	0.56	J
Zinc	(Zn)	21.5	416 to 7400	0.56	J
Boron	(B)	10.0	Declaration	2.81	TMECC
Chlorine	(Cl)	52.3	Declaration	20.0	TMECC
Sodium	(Na)	**ND**	Declaration	281.2	E
Iron	(Fe)	545	Declaration	14.1	E
Manganese	(Mn)	203	Declaration	0.28	J

Particle Size Distribution	Results	Units	Method
< 0.5mm	4.8	percent	F
0.5-1mm	3.8	percent	F
1-2mm	8.0	percent	F
2-4mm	19.9	percent	F
4-8mm	33.9	percent	F
8-16mm	27.7	percent	F
16-25mm	2.0	percent	F
25-50mm	0.0	percent	F
>50mm	0.0	percent	F
Basic Soil Enhancement Properties			
Total (K)	2078	mg/kg	E
Total (P)	249	mg/kg	E
Ammonia (NH4-N)	3.1	mg/kg	A
Nitrate (NO3-N)	2	mg/kg	A
Organic (Org-N)	5935	mg/kg	Calc.
Volatile Matter	14.3	percent dw	D

* "ND" stands for "not detected" which means the result is below the reporting limit.

Method A Rayment & Higginson
D ASTM D1762-84
E EPA3050B/EPA 6010
F ASTM D 2862 Granular
G Butane Activity Surface Area Correlation Based on McLaughlin, Shields, Jagiello, & Thiele's 2012 paper: Analytical Options for Biochar Adsorption and Surface Area
J EPA3050B/EPA 6020

Figure 5.2. Biochar test results for sample B.

The lab will measure your bulk density "as received." Compare the two lab reports. Sample A gives a bulk density of 9.2 pounds per cubic foot (0.15 g/cm^3). Sample B is 13.4 pounds per cubic foot (0.21 g/cm^3). Then look at the particle size distribution. Sample A has 85 percent of the particles larger than 8 millimeters. Sample B has 70.3 percent of the particles smaller than 8 millimeters. The density difference could be due to different feedstocks and production conditions, but particle size has an effect. For a more in-depth explanation of bulk density of biochar, see the article "Weight or Volume for Handling Biochar and Biomass?" by Catherine E. Brewer and Jonah Levine, published in the Biochar Journal online at biochar-journal.org.[3]

Organic Carbon. The total carbon in the biochar that is not bound to calcium or other minerals as part of the ash. The inorganic carbon is reported further down in the data sheet as carbonates. It is important to separate the inorganic carbon for the next test, which is a measure of the carbon stability.

Hydrogen/Carbon. The molar H:C ratio is computed for the organic carbon only, excluding the carbonates. The sample is analyzed to get the total amount of hydrogen. Once the water is gone, all the remaining hydrogen will be bound up in hydrocarbons—tars and oils. This is the volatile fraction that is consumable by microbes and is not as stable as the aromatic carbon of the fixed-carbon fraction. It is easier to accurately measure the hydrogen content of the sample than to determine a volatile fraction by cooking it in an oven. For one thing, when you bake the char again in an oven to try to drive off the volatile fractions, you may end up charring some of that volatile carbon, so the test is not able to accurately measure how effective your biochar process was. Using a lab instrument to determine the total hydrogen content is more accurate. Once we have the total hydrogen and the total organic carbon (excluding the carbonates), we can use the ratio to compare different biochars. The lower the ratio, the more stable the char. The International Biochar Initiative, one of the bodies that sets biochar standards, has determined that a value of 0.7 is the cutoff point for a stable biochar. Any value more than 0.7 indicates that the material is not charred completely enough to be considered biochar. At 0.27 and 0.37, both of our samples are well below the cutoff point.

Total Ash. The total amount of nonorganic minerals as a percentage of the biochar. Keep in mind that if the biomass had been completely burned, it would be 100 percent ash. The ash content of the biochar is partly due to the production conditions, including temperature and oxygen. It is also partly due to the ash or mineral content of the biomass. For instance, wood has about 1 to 2 percent mineral ash, depending on how much bark is included (bark has more ash) and how much dirt was mixed in. Hardwoods have more ash than softwoods. Plant species like grass and corn stover have ash contents ranging from 5 to 10 percent.[4] Both of the reports used primarily wood. Sample A has 8 percent ash and sample B has 4.9 percent ash. These results show that the kiln operators who made sample B did a better job of converting more of the carbon to char rather than carbon dioxide.

Total Nitrogen. Nitrogen is reported here because it can come in different forms and some are bound up in the aromatic carbon structure and are thus harder for plants to access. Ammonia and nitrate forms of nitrogen are reported separately under Basic Soil Enhancement Properties.

pH Value. This property is critical information for knowing how to use biochar in soil. It is related to the ash content, though not always in an obvious way. Biochar is generally more alkaline than soil, and we will discuss ways to neutralize it further on. It is also a property that is easy to measure yourself using inexpensive pH paper.

Electrical Conductivity. This test is also called soluble salts or total dissolved solids. It measures the soluble constituents left in the biochar. Plain biochar won't often have much in the way of salts, but biochar that has been composted or soaked in a nutrient solution will show a high value. If you make biochar from salt-tolerant plants grown in saline soil, this value might test too high for your soil.

Liming and Carbonates. The carbonates are determined so we can subtract the mineral-bound carbon from the organic carbon to determine H:C, as discussed above. Liming value is a measure of the ability of the char to neutralize an acid, and it is presented as a percentage of the liming value of an equivalent amount of calcium carbonate. Some of the liming value in biochar comes from calcium carbonate, and some comes from metal oxides like iron oxide that are also alkaline and provide a liming value.

Butane Activity and Specific Surface Area. Butane activity is a measure of how much butane gas is absorbed by a char sample. Butane will fill the available pore surface of the sample and a pore surface area value measured in square meters per gram of char is reported. Specific surface area is interesting to know, but it does not always correlate exactly to how much of a particular substance the char can hold on to, since the sorption properties of char are very much dependent on the chemistry, pH, and electrical charge of both the char surfaces and the substance you want the char to sorb. Both samples report a fairly high value of specific surface area, between 327 and 335 square meters per gram. Activated carbon used in water filters generally has a specific surface area above 500 square meters per gram. Activated carbon is expensive, and the more surface area it has, the more expensive it is, so biochar can substitute for activated carbon in many applications just by using more of it.

Minerals. These are analyzed and reported for two main reasons: 1) to make sure there are not concerning levels of heavy metals in the biochar and 2) to know if the biochar is a source of important trace minerals such as zinc and boron. These results are easy to read because they give you the threshold levels of the metal that are allowable in soil, where applicable. The code "ND" means "non-detectable."

Basic Soil Enhancement Properties. These are the common fertilizer components of NPK (nitrogen, phosphorous, and potassium), plus volatile matter, which is the portion of the biochar that volatilizes at a low temperature before the char is completely combusted to ash. The volatile matter is not a high precision number but good to know because a biochar with a large amount of easily degraded carbon compounds can cause a microbial "bloom." If you add biochar with lots of volatile matter—usually found in low temperature biochar—to soil, microbes multiply rapidly, feeding on the carbon and consuming whatever nitrogen they can find. If this happens while young plants are trying to grow, the microbes will temporarily rob the plants of nitrogen—not a good result. The same thing happens if you add a lot of sawdust to soil, for instance. It is called nitrogen tie-up or nitrogen immobilization. Luckily it is temporary because once the microbes die, that nitrogen becomes available again as organic nitrogen. This is another reason why it is best to process biochar in a compost pile first before adding to soil.

Bioassays

The test panel done by Control Laboratories is meant to meet the requirements for biochar certification by the International Biochar Initiative; however, these basic tests do not include lab tests for toxic organic compounds that are required for an IBI-certified biochar. Those tests are very expensive, costing thousands of dollars for the full panel on just one sample.

The concern about toxics in biochar is reasonable, but it mostly applies to feedstocks that we don't use for homemade biochar, such as animal manures, construction waste, food-processing residues, sewage sludge, and industrial by-products. For instance, PCBs can end up in biochar only if they are present in the feedstock. To date, only a few tested biochars have shown more than a negligible amount of dioxin and those were biochars made from food waste that had a lot of salt (salt contains chlorine and chlorine is needed to form dioxin).[5] That leaves PAHs, or polyaromatic hydrocarbons. PAH is the term for a huge variety of compounds. In fact, biochar itself is a long-chain polyaromatic hydrocarbon. We are only talking about PAHs of concern, and those are all short-chain compounds like benzene and naphthalene. These PAHs of concern are human carcinogens, but again, biochars that have been tested for them have generally shown very low levels of short-chain PAHs, and they are mostly found in the low-temperature biochars. High temperature flame-cap kilns do not tend to produce PAHs of concern, at least according to the studies and testing done so far.[6]

It is good to be cautious about adding more toxic pollutants to our soil, even if it's not likely to happen. Fortunately, there are some inexpensive bioassays that you can do yourself, which can at least let you know if you have a potential problem with toxic compounds: the germination test and the worm avoidance test.

Germination Test

The goal of germination tests is to determine whether a soil amendment has an effect on seed germination. Seed germination can be inhibited by many kinds of chemicals, including plant enzymes and breakdown products from immature compost, as well as pH that is too acidic or too alkaline. Testing the effect of your biochar on seed germination can let you know that you have a potential problem, but the problem could just be the wrong pH because you applied too much alkaline biochar.

The test is easy to do. All you need are two seedling flats filled with a standard potting soil. In one flat you will mix in some of the biochar at a rate similar to how you want to use it in the field or in potted plants. The other flat is the control. Seed each flat with the same number of seeds. Lettuce (*Lactuca sativa*) is the most widely recommended species to use, due to its sensitivity to the presence of toxins and contaminants in general. You must use many seeds (twenty or more per container) because some seeds might not germinate under any circumstance and to make sure you get a representative sample.

Place the containers in a location where good conditions for seed germination occur. Normal room temperature is most important. Moisten the soil in each container and make sure it doesn't dry out. Use a humidity dome over the flats if you have one, or place a clear plastic bag with some holes punched in it around the flats to prevent them drying out. Check the flats daily for germination. Once significant germination is observed, count the number of seeds that germinated in each container. Don't wait too long to do this, as seedlings might become entangled and will be harder to count. Compare the number of germinated seedlings in the containers with and without biochar, to see if there are differences. If there are no differences, or there are more germinations in the biochar flat, your biochar is good and applied at an appropriate rate. If you are unsure of the result, you can always redo the test. You can also try different biochar application rates.

Worm Avoidance Test

Worms are sensitive creatures and they vote with their feet. In other words, if they don't like the environment they quickly get the heck out of there! Of course, there are lots of things they don't like besides persistent chemical pollutants from industry. For instance, when we tested a large amount of urine-soaked biochar in a worm bin, they said "pee-ew" and crawled right on out of the bin. If you get a bad reaction from a biochar amendment, try a lighter application first before you conclude it is altogether toxic.

For the worm avoidance test, we use the common compost worm species *Eisenia fetida*, commonly known as redworms, although other worms can be used. Use a seedling flat or similar container and make up two soil mixes: one of plain potting soil or a mix of potting soil and clean soil from your farm or garden, and the other of the same soil mix plus the biochar you want to test. Add the biochar at the rate that you would want to use it in your garden or farm field. Make sure the biochar is moist because dry biochar is scratchy and worms will avoid it just for that reason.

Divide the seedling flat in half with a piece of cardboard, and fill one half of the flat with the plain soil and the other half with the biochar treatment. Make sure it is all nice and moist. Then remove the cardboard barrier and place at least ten worms on the dividing line between the two halves. The race is on! Keep the container at normal room temperature and keep it from drying out with a humidity dome or plastic bag with holes in it. After forty-eight hours, replace the cardboard separator (without slicing any worms) and scoop the soil from each half into two separate containers. Count the worms. If the worms have avoided the side of the container where biochar was applied, then the biochar should not be applied to soil without further investigation. Repeating the test more than once or using several replicates will give more conclusive results. You may also want to try different concentrations of your biochar-soil mix, from a 5-percent biochar mix up to a 50-percent biochar mix.

Cation Exchange Capacity

Cations ("cat-ions") are the positively charged ions of molecules. In soil, they are plant nutrients: calcium (Ca^{++}), potassium (K^{+}), magnesium (Mg^{++}), and ammonium (NH_4^{+}). Cations will wash out of soil with the rain if there is nothing to hold on to them. Cation exchange capacity (CEC) is a measure of a soil's ability to retain the important cation nutrients, so it is basically a measure of potential soil fertility. Clay and organic matter have a lot of CEC because they have negatively charged surfaces that can attract and hold on to the positive ions. Sandy soils have low CEC and less ability to hold on to plant nutrients.

Biochar generally has lots of free electrons running around, so it has negatively charged surfaces and can add to CEC when it is mixed into soil. We don't usually test biochar for CEC, however, because it can change a lot when it ages because the surfaces oxidize. It also changes when it is added to soil or compost. When microbes move in to the biochar pores, they deposit organic acids with negative charges onto the surfaces, increasing CEC. On the other hand, as biochar pores get clogged with organic materials and minerals, the surface area may decrease, leaving fewer available sites for cations to bind. It is complicated, but in general, biochar increases the CEC of soil.

Amazingly, biochar can also have anion ("an-ion") exchange capacity (AEC), the ability to attract and hold on to negatively charged ions like nitrate (NO_3^{-}), phosphate (PO_4^{3-}), and sulfate (SO_4^{2-}). There are several mechanisms for this. One source of AEC is metal cations deposited on the biochar surface either during pyrolysis or afterward. This is called a "complexation" reaction that builds up a complex multiatom structure. Each additional compound added to the complex is called a ligand. Another source of AEC is the incorporation of nitrogen in the fused carbon ring structure during pyrolysis. This has a positive charge that can attract anions. AEC functionality is also highly pH dependent.[7]

By the way, I am feeling positively negative about some of the terminology in chemistry, especially the terms for charge. Let's review: A biochar surface has lots of free electrons so it is negatively charged. This attracts cations, which are positively charged molecules. These positively charged cations are now able to attract some negatively charged anions. However, in a battery, the electrode that has a negative charge and attracts positive charges is called the cathode. The positively charged electrode that attracts negative charges is called the anode. That seems confusing to me but perhaps I am easily confused! The most accurately named chemistry term we have discussed is complexation. This stuff is complex.

Biochar and pH

The most important biochar characteristic to know before you add it to soil is where it lies on the pH scale between acidic and alkaline. Flame carbonization produces mostly high-temperature biochar (between 500 and 1000°C), but it is not always consistent. High-temperature biochar has less volatile matter and higher pH (around pH 8 to 10). Low-temperature biochar has more volatile matter that includes organic acids, so it tends to be closer to neutral pH. Always check your biochar pH and your soil pH. Use caution when adding biochar to alkaline soil, but use alkaline biochar to correct low-pH (acidic) soil. Many farmers in Oregon with acidic soils started using a high-carbon ash from local lumber mills because it was cheaper than lime.

A lab-quality pH meter is very expensive and hard to use and maintain. For pH testing that is accurate enough for soil, use a good-quality pH paper. I like the ones from Hydrion. You can get it in different ranges for better accuracy. Here is a list of supplies and a procedure for measuring pH of biochar or any kind of compost, manure, or soil.

Supplies for pH Paper Testing

Quality pH paper from Hydrion
Distilled or deionized water (most bottled water is deionized, that is, pH neutral)
Small jars with lids
Dried and ground-up biochar, compost, and soil samples

Procedure

1. Add 1 tablespoon of sample and 2 tablespoons of water to jar.
2. Cap and shake by hand for one minute.
3. Let sit for 2 to 10 minutes.
4. Dip pH strip into liquid and compare to color chart.

If you have alkaline soil and alkaline biochar, do not despair! There are many ways to lower the pH of biochar and improve it at the same time. The pH issues are one reason that I recommend you always compost biochar before using it. Composting and fermentation are effective in lowering biochar pH because once microbes move in and start eating and excreting, they deposit organic acids on the pore surfaces of the biochar, neutralizing some of the alkalinity from the ash.

Other ways to begin to reduce alkalinity include simply rinsing the biochar in water or leaving it out in the rain for a season. You can also add an acidifier like vinegar, citric acid, or sulfur. Wood vinegar is an especially valuable acidifier that we will talk about in chapter 6.

Particle Size

Biochar from the kiln is usually a mix of big chunks, small granular particles, and fine, micron-size particles. When do you need to crush the chunks into smaller sizes and when can you just use the biochar as is? It helps to know that different particle sizes do different jobs in your compost and soil.

Finer particles can be consumed by worms or other soil organisms. Finer particles also have internal pores that are more accessible by channels to the surface. That means more available surface area to adsorb nutrients and provide habitat for microbes. To a bacterium, a tiny char particle is a whole city.

Larger particles have benefits for bulking and aeration. If you have a deep raised bed, a large pot, or a planting hole for a tree or shrub, fill the bottom with biochar chunks for aeration. I like to add some bokashi as well. Bokashi is an anaerobic ferment that I will talk about extensively beginning in chapter 8. It is like naturally fermented pickles, and it can displace less beneficial anaerobes in the deep, dank places in your soil.

Large chunks of biochar are also fine for use in animal barns and chicken yards. The animals will stomp on or peck at the char chunks and soon they will fracture and get incorporated in the manure, building a nice compost right in the barn or corral. Chunky biochar also makes a good mulch around perennials like fruit trees and berry canes. A thick layer of several inches (about 5 cm) of char chunks and fines will help suppress weeds, especially grass.

For adding to compost piles, a mix of particle sizes is probably best. A half inch (1 cm) and smaller is a good goal. Put down a layer of chunky biochar on the ground and build your compost pile on top of that in layers with finer biochar particles.

For mixing with soil blends and fertilizers, I prefer a quarter inch (6 mm) or less, a granular size that mixes easily with organic fertilizers like seed meals, feather meal, and crustacean meal. Granular-sized biochar is also very useful in a bokashi food composting bucket.

Methods to crush biochar include using roller mills, hammer mills, and other devices. My friend Bob Wells of New England Biochar uses a shredder/crusher machine designed for recycling plastics to quickly and efficiently chew through the large volumes of biochar that he makes. The easiest way to get started is to lay a tarp out in your driveway, spread biochar on it, and then drive over it back and forth about twenty times with a truck or a tractor. You can put another tarp on top of it to keep the biochar out of your tires. If you have included used lumber with nails in your kiln, do not forget to remove the nails first! An inexpensive magnet sweeper from the hardware store is the tool for this job. Or find another way to crush that biochar.

I ended up purchasing an electric hammermill from PelletMasters.com to do the job of crushing and learned a few things about the best way to use it. A hammermill is a series of hammers mounted on a rotary shaft that spins and slams into the material, hitting it multiple times until it can pass through a

Figure 5.3. Processing biochar with the hammermill (*top*). The innards of the machine showing the swinging hammers and the oversized screen that I use (*bottom*).

screen. Biochar needs to be moist before processing or it will produce a huge black dust cloud. However, moist biochar is sticky and can clog the screen. Use a large screen size to prevent clogging. I swapped out the 6-millimeter screen it came with for a 10-millimeter screen.

If the char is too wet, it won't go through the screen at all, so you need to find the sweet spot of moisture to be successful. Start by spreading moist biochar out on a tarp in the sun for a few days to dry. Try it, and if it is still too wet, it will clog the hammermill screen quickly. The best option is to wait and let it dry more. If you get it too dry and it makes a dust cloud, you can lightly spray with water and get the moisture content to the sweet spot. It does not take much water to keep the dust under control. It may still clog a bit, but the screen is pretty easy to get to and clean if needed. I find that I can process about one cubic yard (0.8 m^3) of biochar per hour, stopping to unclog the screen every 20 minutes or so. It is a slow, dirty job, but good char washes off easily! Do wear a dust mask and eye protection. Ear protection from the loud machine is a good idea, too.

Water-Holding Capacity

Biochar can hold more than its own weight in water, although the water-holding capacity of biochar varies widely with feedstock, production temperature, and quenching method. Biochar made at low temperature, like barbecue lump charcoal, is generally hydrophobic, meaning that water will bead up on the surface and take some time to be absorbed. That happens because low temperature biochar has retained a lot of the tars and oils, or volatile matter, that was in the wood.

Biochar made in a flame-cap kiln is high-temperature biochar. Not only is most of the volatile matter gone but, because the char is exposed to some oxygen in the process, the pore surfaces are somewhat oxidized, which makes the char less hydrophobic. If it is also quenched with water, then it will be very friendly to water (hydrophilic) and easy to wet. Furthermore, it will have a lot of pore volume to hold water like a sponge.

But biochar pore volume is not the whole story with water-holding capacity in soil or even in a bucket of pure biochar. Particle size and shape also play a role because a lot of water is held in-between particles. This is called "inter-particle" capacity. The water-holding capacity inside the biochar pores is called "intra-particle" capacity. A review of many different kinds of biochar found that biochar that was hydrophilic and that had mixed particle sizes with irregular, rough surfaced shapes could hold up to four times its weight in water. Hydrophobic biochar with small particles and smooth surfaces held much less water.[8]

Biochar impacts soil water-holding capacity differently depending on the soil. Biochar has a big impact on coarse, sandy soil that is free draining, by filling in the channels between coarse soil particles with the electric carbon sponge that attracts water. In a heavy clay soil, biochar can reduce compaction and open up the inter-particle channels, allowing better drainage.

You can test the power of biochar to hold water in your soil very easily. Make two soil mixes, one with your regular soil and one with a measure of biochar. Or make several mixes with different proportions of biochar. Fill a few pots and grow a tomato plant in each one. When the plants get big enough to start flowering, quit watering all of the pots and see who survives your drought and how much longer it lives. No doubt the plant without any biochar will wilt first.

Temperature of Formation and Carbon Stability

The temperature at which biochar is made has a determining impact on its characteristics, as we saw in table 5.1 (page 90), which showed a fractional analysis of ponderosa pine and grass baked at different temperatures. Without doing a lab test, we can make some good assumptions about the fractions of fixed carbon, volatile matter, and ash, just by knowing the temperature of formation of the char.

Another reason to know the temperature of formation of biochar is because new possibilities are opening up for carbon removal payments for making biochar. Some carbon accounting protocols require that a minimum temperature of 600 to 650°C is reached because biochar formed at that high temperature is assumed to be highly recalcitrant and not degradable for hundreds to thousands of years.

Temperature can be measured in the field, where we are making biochar, but that is not very easy to do without a special temperature probe that reaches into the center of the kiln. An infrared thermometer can measure temperature of the metal kiln walls, but that is not the same as the temperature where char is forming. Aiming the sensor at the flames does not work because the flame luminosity interferes with the infrared signal.

The adiabatic (without any heat loss by heat transfer) flame temperature of burning wood is estimated to be close to that of propane, or 1,977°C. Biochar production in a flame-cap kiln is closely coupled with the flame, without heat transfer losses by conduction through a metal wall, as in retort pyrolysis. Therefore, we would expect that the temperature of production would be high as long as a flame is maintained during the process. A survey of chars using Raman spectroscopy reported that a biochar sample from a flame-cap kiln (one of my cone kilns) was among the three samples with the highest apparent temperature of char formation, in the range of 900°C.[9]

I think we can safely say the biochar made in a Ring of Fire kiln or other flame-cap kiln is high-temperature biochar with a high percentage of fixed versus volatile carbon. Going forward, for purposes of verification that biochar was produced at temperatures above 600°C in a flame-cap kiln, it is good enough to just monitor the flame during the production process to confirm that biochar is produced in a flaming and not a smoldering combustion process. If you simply take a series of photographs and short videos during the production process in a flame-cap kiln, that should serve to verify that the carbon will be highly recalcitrant and eligible for carbon payments, as discussed in chapter 13.

How to Store Biochar

The best way to store biochar is with moisture. Dry biochar can be a dust hazard, and it can also be a fire hazard if it is close to an ignition source. Some forms of biochar can even ignite on their own in a process called self-heating. This can happen even with biochar that is slightly damp. The danger of self-heating is mostly from biochar made at low temperature that still contains volatile oils. Researchers in Britain made a series of biochar materials from wood and straw at different temperatures and then measured their propensity to self-heat. They found that biochar was most prone to self-heating when it was produced at 450°C. Biochar made at 600°C was less reactive than the original feedstock.[10]

We can conclude from this that biochar made by flame carbonization at high temperature will be less likely to ignite in storage. However, it is still a good idea to be careful when storing biochar. Keep it in smaller piles and sacks, and keep some space between the sacks. The best solution is to just use it as quickly as possible in compost, manure management, or soil. Once it is mixed with other ingredients in a moist compost pile, there is little danger of ignition, although sometimes compost itself can ignite if it is in huge piles with lots of wood chips. This is not likely to happen in a small home-garden compost pile.

CHAPTER SIX

Transforming Smoke into Useful Chemicals

When we make biochar, we are making more than just carbon. Taking apart wood with fire, our "pyro-lysis" releases vapors that can form smoke. Are there any beneficial uses for smoke or is it just pollution? The answer is that smoke contains chemicals that have many potential uses.

Smoke is only one part of the emissions from pyrolysis. Carbon dioxide, carbon monoxide, and methane are not visible emissions but they have other, well-known problems, especially their role as greenhouse gases. The best way to reduce carbon dioxide emissions is to convert more of the fuel to biochar. The best way to reduce emissions of carbon monoxide and methane is to burn them in the flame. The same goes for the stuff that makes smoke.

Smoke is a combination of unburned particulates and vapors. The vapors are mostly water vapor and volatile organic compounds (VOCs), a term that can include dozens of elements and compounds. When making biochar, we try not to let smoke get away from us. When it is not possible to burn the smoke, there are ways to condense and cool it into a liquid called wood vinegar (WV) or pyroligneous acid (PA). I will mostly use the term wood vinegar here, but there is increasing interest from the scientific community in this substance, and many of the research papers refer to wood vinegar as PA.

Charcoal makers in Asia have long known of wood vinegar and its use as a plant growth stimulant. In the United States and Europe, wood vinegar was used as a mordant, or fixer, for fabric dyes, and sometimes sold as a substitute for apple cider vinegar, but farmers were seemingly unaware of its agricultural uses.

Wood vinegar is composed of 80 to 90 percent water and 10 to 20 percent organic compounds. Acetic acid is the largest single organic component at about 5 percent. Wood vinegar also contains various kinds of phenols, ketones, alcohols, esters, aldehydes, and furans, including many that act as plant enzymes. Wood vinegar is used in agricultural crop production to accelerate compost, stimulate germination and plant growth, and also for soil disinfection and the control of weeds, disease, and pests.

You might wonder how the same compound can act as both a pesticide and a growth stimulant. The answer is in the dilution rate. The Government of Thailand promotes wood vinegar to farmers as a sustainability practice and advises the following dilution rates:[1]

- Kill pests by applying a 1:20 ratio to the soil
- Kill weeds by applying a 1:50 ratio to the soil
- Prevent stems and roots from rotting by applying a 1:100 ratio
- Prevent pests and mold while accelerating plant growth by applying a 1:200 ratio to the soil
- Prevent plant lice (aphids) by applying a 1:400 ratio to the plant
- Enhance fruit growth by applying a 1:500 ratio to forming fruit

The pest- and disease-control properties of wood vinegar may be primarily an effect of the acetic acid (pH 3.0) and phenol compounds, but the complex mixture of organic acids and alcohols contributes in unknown ways to the effects observed. For example, because wood vinegar has small quantities of various kinds of alcohols and similar chemicals, it is easily absorbed into plants. It also dissolves other agricultural chemicals, making them easier to absorb, according to the wood vinegar handbooks I discuss further on, although no specific studies are readily available to corroborate this.

But what about the plant growth stimulation claims? Plants use hormones to regulate reproduction and growth. Condensed smoke from vegetation fires produces karrikins, a group of plant growth hormones that promote seed germination and rapid seedling growth. Any plant species that evolved with fire in its environment is likely to respond well to karrikins. When a plant senses this chemical in its environment, it gets the message: "There has been a fire." The plant interprets this message to mean: "A fire has killed everybody else. I am the only one left. If I ramp up my metabolism right now, I won't have much competition, so go for it!"

Karrikins and other growth hormones also play a role in the metabolism of microbes, which could account for additional growth-stimulation effects and be responsible for speedier production of compost. However, researchers have found that if you add too much, wood vinegar can inhibit growth, so pay attention to the recommended dilutions.[2]

Overall, refined wood vinegar seems to be safe for the environment and workers. A Finnish study of wood vinegar found it to be of low environmental risk because it is rapidly degraded through microbial activity, which also means that to be effective as a pest control, it will have to be reapplied on a regular basis.[3] Wood vinegar supports the movement away from petrochemical-based synthetic agricultural chemicals that are persistent organic pollutants, because wood vinegar works effectively with a range of organic inputs that are more sustainable and compatible with natural life cycles.

Wood Vinegar Australia makes a wood vinegar product called PyroAg and has conducted numerous trials in various applications. One of their customers reported using the product to improve the macadamia tree's resistance to husk spot disease, which causes early nut drop. According to Wood Vinegar Australia's website, macadamia farmer Guido Conte says, "What I noticed was that by foliar spraying the PyroAg, it helped to mature the nut quicker and therefore immaturity was greatly decreased. It seems to be a very good tonic for the macadamia trees."

Researchers, mostly in Asia, have documented significant yield increases in vegetable production using diluted applications of wood vinegar, including a 20 percent increase in lettuce and cucumber,[4] and significantly improved plant growth, fruit weight, fruit diameter, and sweetness of melons.[5] Looking to increase rice seed germination under dry conditions, researchers primed seed with 1:50 diluted wood vinegar mixed with biochar. The combined treatment improved germination by 67 percent and increased grain yield by 20 percent.[6]

Mixing wood vinegar with charcoal can have synergistic effects. Korean scientists found that fruit size, sugar content, and marketability of sweet peppers were most enhanced using a mixed treatment of wood vinegar and charcoal powder.[7]

Wood vinegar seems to help plants survive various stressful conditions. It improved the growth of rapeseed seedlings under freezing temperatures, most notably by enhancing water use efficiency.[8] In a saline soil, wood vinegar as a foliar spray improved the growth of rapeseed plants by reducing their salt uptake and improving their ability to photosynthesize.[9]

Many studies have looked at wood vinegar applied as a fungicide. It completely eliminated alternaria blotch of apple with a 1:32 wood vinegar-to-water dilution and decreased fusarium infections in wheat by 66 percent when sprayed at a 1:200 dilution rate.[10] This level of control was similar to that attained by commercial chemical fungicides.[11]

Wood vinegar works as a pesticide and there is also the interesting possibility of using wood vinegar to enhance the action of other pesticide chemicals, including essential oils. A coffee-berry pest that infests berries in the field and also in storage was controlled with a combination of wood vinegar and clove oil.[12] A similar combination of wood vinegar and garlic extract could be effective on pests and fungal infections. Wood vinegar can deter many insects, which is one reason to be cautious about spraying it during flowering. Crops will suffer if wood vinegar drives away their pollinators.

Wood vinegar is also very useful in ferments and composts. At the right dilution rate, wood vinegar stimulates microbes. If too strong, it will kill bacteria in compost but promote fungal degraders. Chinese researchers added wood vinegar to silage—a lactic acid bacteria ferment—and found that wood vinegar increased the lactic acid content of the silage and decreased gas production, especially that of CO_2. Silage is a growing source of CO_2 emissions as it is a major component of animal feed.[13]

Some of the most interesting research has identified the hormetic nature of wood vinegar application. Hormesis in biology is the recognition that a small dose of a substance can elicit a beneficial result where a larger dose would be toxic or inhibitory. Wood vinegar showed this property in many studies.[14]

The dose makes the poison, but another factor to consider is that wood vinegars are not all alike. Generally, the pyroligneous acids distilled from most tree species have more heavy tar fractions than those made from bamboo. These heavier fractions may inhibit germination, and even too much of the lighter fractions can suppress growth. Softwoods have more terpenes than hardwoods, and the terpenes should not be applied directly to plant foliage. If you purchase wood vinegar, it should have a clear, light amber color and not be oily. If you make it yourself (more on that below), you need to refine the product by separating out the tar and terpenes. Terpenes are found in the light oily fraction on the top of a container of unrefined pyrolysis liquid.

How to Use Wood Vinegar

Interest in wood vinegar is increasing, and we are learning more about wood vinegar and its uses. It is difficult to draw solid conclusions, because not only is wood vinegar chemistry widely variable, but application rates make a huge difference in the research results and conclusions. Research into wood vinegar continues, but for practical guidelines today, the best resources I have found are two manuals from Japan. The first is called *The Uses of Wood Vinegar and Charcoal in Agriculture* by Hirowaka Tsuyoshi, published in 1994.[15] I have only a paper copy of this document and have not been able to find it online. The second manual is a seventy-five-page document by Masaki Yokomori, published by the Japan International Cooperation Agency (JICA) and based on a three-year agriculture cooperation project with the farmers in Benguet province, Philippines: *Farmers in Benguet Practice Savers Technology*.[16] You can enter the title as a search term to find and download your own copy.

I have used both of these manuals to guide experiments with wood vinegar in my own vegetable garden. My friend Michael Wittman, who distributes all

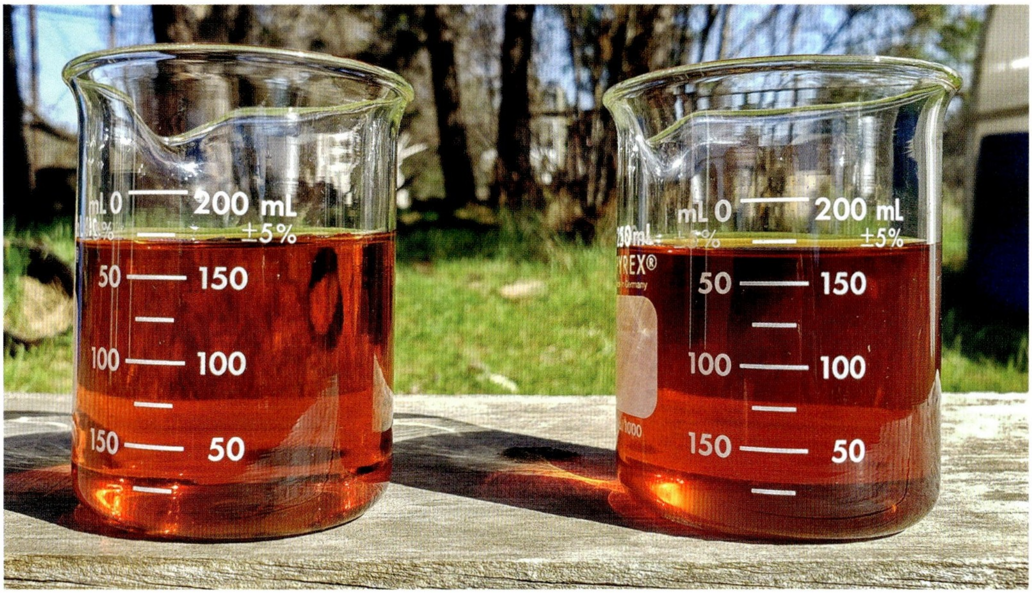

Figure 6.1. Two different kinds of wood vinegar available commercially. The vinegar on the left is made from hardwood, and the vinegar on the right is made from bamboo.

kinds of interesting biochar products through his online store, Blue Sky Biochar, has supplied me with plenty of wood vinegar. When choosing wood vinegar products, Michael looks for a clear amber color with no dark tones. The vinegar should not have an oily feel to it, and the pH should be between 2.7 and 3.1.

Here's a brief synopsis of the basic guidelines for using wood vinegar, drawn from both of the Japanese manuals and my own experiments.

WOOD VINEGAR–SOAKED BIOCHAR

Mixing wood vinegar with biochar has many benefits, and you need to use a relatively large amount of wood vinegar in this application. Mix one part of wood vinegar in five parts of water and sprinkle it slowly and evenly on a pile of dry, fine-grained biochar, letting it soak in well. It should get just wet enough to pass the squeeze test: Take a handful and squeeze it. Open your hand and see if it crumbles. The moisture is correct if the mixture stays in a clump when you open up your hand.

Let the wood vinegar–charged biochar dry and use it immediately in soil or store in an airtight container. If using directly in soil, apply at a rate of about 200 pounds (91 kg) over a quarter acre, or 2 pounds (1 kg) per 100 square feet (9 m^2). Do this once every season to help condition soil and promote soil microbial life.

I have used this technique, not to charge biochar for soil, but to change the pH for use in composting and fermentation. I was able to reduce the pH of biochar from 8.5 to 5.5. I have used this acidic biochar in manure composting to help control ammonia. If you have very alkaline soil, wood vinegar–soaked biochar might be your best option for successfully using biochar.

WOOD VINEGAR SOIL DRENCH

This is the main way I use wood vinegar, as part of liquid fertilizer drenches that I apply throughout the growing season. Since I already have a lot of biochar in my soil, wood vinegar and other nutrients can soak into it and recharge it with nutrients. A stronger solution of wood vinegar is appropriate for use in soil. A dilution of 1:50 (WV to water) is a good place to start if you already have plenty of organic matter in your soil. High mineral content alkaline soils may react with acidic wood vinegar to produce carbon monoxide and carbon dioxide, so if applying wood vinegar as a soil drench to such soils, it is advisable to wait a week or two before planting to let the reactions finish.

Hirowaka Tsuyoshi says that wood vinegar soil drenches increase plant growth, promoting bacteria such as bacilli, actinomycetes, and Trichoderma. These bacteria help control rhizoctonia, pythium, and other soil pathogens. Since I have been using plenty of biochar compost and wood vinegar in my garden for years now, I have not seen any of these diseases. Nor have I had clubroot nematodes, which Tsuyoshi says are killed by wood vinegar.

I have occasionally used wood vinegar directly in soil at a 1:3 (WV to water) dilution to kill pests. One spring, I saw that cutworms had attacked the stems of my broccoli plants. Two plants were already badly affected before I noticed. I soaked the stems and roots with wood vinegar solution and saved all the plants but the two that had already keeled over.

FOLIAR GROWTH SPRAY

One of the most immediately noticeable impacts of wood vinegar in the garden is from the foliar growth spray I have been using. In my 3-gallon sprayer, I mix the following ingredients:

3 gallons (11 L) water
4 tablespoons (60 ml) wood vinegar
1 cup (240 ml) EM-1 microbial inoculant (see chapter 8 for more on this)

I spray this in the greenhouse in the spring to freshen things up and give the seedlings a boost. I spray it on growing plants in the garden and within a day or two I can see that the leaves are greener and thicker. I spray it on fruits and berries that are ripening—supposedly it makes them sweeter. Recently, I noticed that the leaves on the grapes were looking quite pale so I gave them a good spray and within a few days they were dark green again. Hirowaka Tsuyoshi says that the improvement in photosynthesis is due to the esters in wood vinegar that act as coenzymes to stimulate photosynthesis and growth.

WOOD VINEGAR FUNGICIDE AND PESTICIDE SPRAY

I have tried wood vinegar spray as a fungicide, using a dilution of 1:200. This works out to about 60 milliliters (¼ cup) of wood vinegar in my 3-gallon sprayer. I have found it to be very effective on gray mold (*Botrytis cinerea*) on strawberry and against powdery mildew. It prevents damping off of seedlings in the greenhouse, and it worked to kill a rust fungus that attacked my onions and garlic one year.

I use the same dilution as a pesticide. I have had success eliminating aphids and flea beetles in the garden, and whiteflies in the greenhouse. It does not seem to do much to leaf miners, squash bugs, Colorado potato beetles, or cabbage butterfly eggs and larvae. I would like to try making a wood vinegar and essential oil extraction to make a better pesticide for the bigger bugs. The Benguet farmers manual recommends soaking 100 grams of chili, neem, garlic, ginger, or bee propolis in one liter of wood vinegar for two to three months, stirring weekly. Then dilute it by 100 to 500 times with water to make a spray.

Some of these extracts will also be effective as fungicides. Ag consultant Mike the Worm Guy (aka Michael Flynn), who corresponds regularly on the biochar forum at biochar.groups.io, has told us about a spray he developed for large farms that grow leafy greens. The spray includes wood vinegar, garlic, and vermicompost extract to control plant pests and diseases, especially powdery mildew and downy mildew. You can reach Mike Flynn through his company, GreenQuest.

The one problem that wood vinegar did not help me with was early blight (alternaria) on my tomatoes. This problem has plagued me every year, and nothing seems to help. But I reached out to the biochar forum and got some good suggestions. Mike the Worm Guy came through again with a spray recipe:

1. For each gallon (3.8 L) of water, add 1 percent by weight (38 grams) of baking soda (sodium bicarbonate) or potassium bicarbonate. Stir well and dissolve.
2. Add 0.5 percent by volume (.64 ounces or 19 milliliters) of wood vinegar and stir.
3. Add .05 percent by weight (2 grams) of diatomaceous earth and stir.
4. Spray and wet plants thoroughly. Pay attention to spray the stem fully.

I asked Mike how this spray works because I was perplexed by the combination of acidic and alkaline ingredients. Here is his answer:

> This recipe is a layered recipe of partial solutions that together can keep the blight at bay. The partial solutions include alkaline pH from the baking soda (sodium bicarbonate) and the making of sodium acetate by adding 0.5 percent wood vinegar to the baking soda solution. The baking soda alkalinity keeps the pH up. The organic acids from the wood vinegar combine

> with the sodium in the baking soda thus making sodium organic acid salts. Sodium organic acid salts can have an antimicrobial effect. The diatomaceous earth (DE) has biogenic silica. The silica in the DE is less reactive than using potassium silicate (potassium silicates do not always play well with others). You can use potassium bicarbonate in place of sodium bicarbonate if you want the potassium. Also keep in mind there are three parts to any effective spray approach: a good recipe, application frequency, and coverage.

Silica helps a plant fight foliar fungal diseases like alternaria, so in addition to Mike's spray, I used a potassium silicate soil drench on all the tomato plants (about fifty). As I am writing this in mid-July, the tomatoes look amazingly good. A few plants have a few spots, so I will get out there and spray one more time, but compared to last year, my tomatoes are extremely healthy and I am starting to wonder how on earth we will get them all canned!

To use wood vinegar with chemical pesticides, the Benguet farmers manual says to reduce the recommended dosage in half and mix with wood vinegar diluted by 50 to 100 times with water. However, this will only be effective with pesticides that are acidic. Copper and lime-based fungicides should not be mixed with wood vinegar. For the farmers of Benguet, reducing pesticide use is an important cost savings.

How to Make Wood Vinegar

Bob Wells is the owner and chief designer of biochar systems at New England Biochar in Eastham, Massachusetts. Bob took an improved brick charcoal kiln design called the Adam retort (created by Chris Adam for use by charcoal makers in Africa), and redesigned it to make a moveable, steel-fabricated unit with process controls for use on small farms. Bob was driven to make the kiln as clean and efficient as possible, and that first charge of wet gas that comes out of the retort and will not burn was a problem to solve. He decided to condense the gas before he ever heard of wood vinegar, but now the wood vinegar condenser is an integral part of the New England Biochar retort that he builds as custom orders for clients. Many of his clients are just as interested in the wood vinegar as they are in the biochar it makes.

A farmer himself, Bob Wells has seen 20 percent yield improvements using wood vinegar. He also has been using wood vinegar successfully in his field of blueberries to deter bud-munching gypsy moths. Now his neighbors are trying it, too. "Winter moths are killing trees here on Cape Cod, and some of the towns are starting experiments to see if wood vinegar sprays can save our trees," Wells says.

Bob's retorts are state-of-the-art units perfect for small farms or someone who wants to go into the business of making biochar and wood vinegar. There

Figure 6.2. A. This New England Biochar retort was installed at Swallow Valley Farm in Sonoma County, California. B. The condensed smoke is collected in a tank. C. Bob Wells flares the light gases that do not condense (mostly methane and carbon monoxide) as a way to control the temperature in the kiln when it gets too hot.

are also some simpler technologies that are being developed for small-scale production of wood vinegar. The Benguet farmers manual includes a diagram and some photos of a kiln—made from a 55-gallon drum—that makes biochar and wood vinegar. The drum is laid horizontally with a hole in one end for a firebox and a hole at the other end for a chimney. There are several different ways to make a condenser. The traditional way is to use a long bamboo pipe, long enough to cool and condense the smoke, and slanted at 30 degrees so that the condensate drains back to a slot in the pipe where the wood vinegar drips out and is collected. The other way to design a condenser is to use a metal flue pipe surrounded by a water jacket to cool and condense the smoke.

Mr. Arun Waikham, along with partners Boonsong Thansrithong and Patrick Trail at the ECHO Asia Small Farm Resource Center in Chiang Mai, Thailand, have created the innovative "Do-All" biochar burner design.[17] This charcoal retort

Figure 6.3. Production of charcoal and raw wood vinegar after a single burn with the "Do-All" biochar burner at the ECHO Asia Small Farm Resource Center, Chiang Mai, Thailand. Photos courtesy of ECHO

includes an integrated smoke condenser for collecting wood vinegar while making biochar at the same time. The retort is designed to be fired by placing it on a wood-fired stove or other external heat source. The article at ECHOCommunity.org includes detailed dimensioned drawings for anyone to use to make their own. The authors report that the unit can make about 10 kilograms (22 lbs) of biochar and 10 liters (10.6 qts) of condensate in four hours of firing.

My plan is to make something like this, but it will be an autothermal kiln (like the traditional charcoal kilns discussed in chapter 3) and not an externally heated retort like the ECHO retort. I will start with a drum smoker kit. The BBQ culture in the United States is populated by passionate grillers and smokers who go to great lengths to perfect their craft. Making a meat smoker out of a drum is a fun project and you will find many online videos detailing every aspect of it. It struck me that this basic drum smoker is exactly what I need to make the base of my wood vinegar kiln. All I need to do is design the smoke condenser to go with it.

To build your own wood vinegar kiln, take a look at Ugly Drum Smokers, a website that caters to the home BBQ enthusiast: udsparts.com. There you can purchase all the accessories you need to smoke a turkey, including air dampers, grates, baffles, baskets, and thermometers. You might as well make a dual-purpose kiln that can do the barbecue thing and then switch to smoke-condensing mode when the cooking's done. After all, that is what barbecue is all about, condensing smoke on cooking meat to enhance flavors. Wood vinegar is just liquid smoke.

Smoking food is an ancient practice and an important method of food preservation. The smoke condensates are antibacterial. One of the uses of wood vinegar being investigated today is for food preservation for the food processing industry.

There is a real art to making high-quality wood vinegar and not just a bucket of toxic tar. Hirowaka Tsuyoshi has several recommendations for operating a kiln to produce high-quality wood vinegar that I summarize here:

Choose your wood. It should be neither too wet nor too dry. Green wood is usually about 50 percent moisture, so if you cut the wood and let it dry for about three weeks out of the rain, it should be about 35 percent moisture, which is just right for making wood vinegar. Any log bigger than fist-size should be split. If using grapevines or fruit tree prunings, it should be drier, about 20 percent moisture, because these materials have a higher ash content and are harder to carbonize.

Choose your materials wisely. Use stainless steel, clay, or bamboo for your condenser pipes because iron pipes will oxidize. Store the wood vinegar in glass or acid-resistant plastic. A drum kiln can produce about 15 kilograms (33 lbs) of charcoal and 3 liters (3.2 qts) of refined wood vinegar from one firing of 60 kilograms (132 lbs) of wood.

Collect the smoke at the right time. White smoke during the initial part of the burn is mostly water. Start condensing when it turns yellow. When the smoke turns a purple-blue at the end of the burn, stop, because the high temperature at the end of the burn produces mostly tar.

Do not let the kiln get too hot. Temperatures above 425°C in an autothermal kiln with restricted air (as opposed to a flame-cap kiln or gasifier) will form more tar, including PAH compounds with benzopyrene.

Refine the liquid before use. This step is to remove the heavy tar fraction and the light oil fraction. Only the water-soluble vinegar should be used directly on plants. The best way to do this is by letting the liquid settle into fractions for at least three months. The liquid continues to react for some time as the formaldehydes form resinous tar by polymerizing with phenols. After three months, you can remove the thin layer of oil on the top with a syringe and siphon the wood vinegar off of the tar layer at the bottom. If the wood vinegar color is still dark with tar particles, add some powdered biochar and let it sit for more time. The tar particles will stick to the powdered char and sink to the bottom.

Tsuyoshi says that the light oil can be mixed with the tar and applied to compost where it will decompose quickly, feeding microbes. Bob Wells takes the light oil and puts it in a jar with a wick to make an oil lamp.

CHAPTER SEVEN

Ashes and the Value of Minerals in Biochar

Smoke leaves the biochar behind, but ash stays, becoming an integral part of the char. How much ash remains depends on the process and also the feedstocks. Biochar can be made from many things, including bones. Bones are much lower in carbon and much higher in ash than wood and other plants, so bone char is a good example of what biochar with a lot of ash minerals can do.

Charring the Bones

Bone char only has about 10 percent carbon char but is a valuable fertilizer containing phosphorus and calcium. A commercial bone char rating is NPK 0-16-0 with 16 percent available phosphate and 32 percent total phosphate plus 33 percent calcium. It is unclear whether bone char phosphorus is either more or less available to plants than the phosphorus in fine-ground bonemeal, but it is very clear to me that it is much easier to crunch up brittle bone char for fertilizer than to grind the tough old bones.

Bone char is best made in a retort process that can be as simple as tossing a paint can full of chicken bones on a flame-cap kiln. I have made lots of chicken bone char this way. I pick the bones out of the finished compost when I am screening it for use in the garden. Then I pack them into a clean 1-gallon paint can with a lid. I poke a few holes in the lid to let the gas escape. After the bones are charred I pulverize them with a mortar and pestle. Bone char is especially valuable as a fertilizer for tomatoes, which need plenty of calcium and phosphorus to avoid blossom end rot.

A paint can works for normal household food waste, but if you have more and bigger bones to char, you will want a dedicated retort.

Figure 7.1. Cattle bones before and after charring. Photos courtesy of Callicrate Cattle Co.

Mike Callicrate is a Colorado rancher who found himself in need of a way to char lots of cattle bones. Normally, in our food system, ranchers grow the beef and then send the cows off to feedlots and slaughterhouses for finishing, slaughtering, packaging, and distribution of the meat. But Mike did not like the new trends for consolidation in the industry:

> As a rancher, I've experienced first-hand how damaging consolidation can be: monopolistic beef packers and food retailers have squeezed the livelihood out of the daily care for land and livestock. With four meat packers now controlling nearly all beef processing, and a handful of large retailers partnering with the same meat packers to control access to the marketplace, the producer share of the food dollar has plummeted to all-time lows.

Mike decided to take back control and put together a mobile slaughter house and refrigeration system to process his own cattle and sell the meat directly to consumers. He composts the waste with biochar, except for the bones. After the bones are boiled for soup stock, they go into the retort to make bone char.

Mike uses the Exeter Retort, a wood-fired retort with a capacity of 1.7 cubic meters (2.2 yd^3), to make the bone char that he uses on the ranch. Besides helping his own bottom line, he is aware of the need to conserve and make the best use he can of phosphorus, because "there is only a limited amount of phosphorus left in the world and it is all under the control of four big mining companies." For self-reliance, we need to husband supplies of phosphorus and stop wasting the bones. Learn more about Mike's bone char and how to purchase his meat products at the CallicrateCattleCo.com website.

Ashes to Ashes

Ashes are nutrients that plants need. If you burn crop waste and return it to the soil, you have replaced a lot of the soil minerals that were used by the crop, reducing the need for mineral inputs to keep growing that crop on the same soil. While crop wastes are not as easy to pyrolyze as wood, their ash can make better fertilizer.

There is no better example of the circular economy than returning nutrients to soil, whether as ash, char, bones, manure, or crop residues. The practice of field burning is the quick and dirty way to do this. An alternative is to let fields fallow for a year in between crops so that stubble and stover can slowly decompose and become part of the soil again, but farmers cannot afford that, which is why they burn fields. Burning releases the ash, making its nutrients immediately available to the next crop.

Field burning is frowned upon in the United States these days, so crop waste is either hauled off the field for various uses, or it is left in place to decompose, but there is a lot of interest in making biochar from straw, corn stover, and other residues. High-ash-content straw and stover biochar materials have fertilizer value.

The International Biochar Initiative has a classification system for biochar materials based on four attributes: carbon storage value (fixed carbon), fertilizer content (mineral nutrients), liming value (calcium carbonate equivalent), and particle size. The fertilizer value is based on the requirements of a maize crop, which is appropriate because corn stover is the most likely kind of crop waste to be carbonized.

Corncobs, as we saw in chapter 4, are fairly easy to carbonize in a 55-gallon drum TLUD. Rice hulls are easy to burn in a small fan-assisted TLUD cookstove. The high silica content of rice hulls facilitates the formation of char.

Figure 7.2. Hemp stalk biochar added to the compost pile.

Other crop residues, like straw, can be processed and pelletized for use in a TLUD cookstove. Some high-ash crop residue feedstocks that we can use in flame-cap kilns are items like hemp stalks, berry canes, grape prunings, and orchard prunings. These feedstocks are a bit more on the woody side, but they have more ash content than wood. When using these feedstocks in a flame-cap kiln or TLUD, they must be very dry or they won't burn well at all. This can be a consideration when fitting this task into the farm routine.

I tried to help my neighbor char hemp stalks in a Ring of Fire kiln, but the material was too wet. We made some char for his compost pile, but it was at the expense of a lot of smoke. Drying the hemp stalks first is essential, but that would require a covered space to keep the piles out of the rain. The need for covered storage could be avoided if this farm could char at the right moment in the fall after the green stalks have dried a bit, but before the rains start. That timing does not work, though, because at that time the farm is in the middle of harvest and there is no labor available to do the job. Considerations like this open the way for contractors to offer biochar production as a service to farms, when that makes economic sense.

Biochar Mineral Complexes

Biochar properties change with production conditions (oxygen, temperature, heating rate), aging (moisture content and oxidation of surfaces), and after addition to soil or compost, as dissolved nutrients and minerals are absorbed. It is also possible to change the properties of biochar by adding minerals during the production process to create what are called biochar-mineral complexes. Typically, this is done by cooking the biomass with clay minerals, iron compounds, or even organics like manure.[1]

Biochar researcher Dr. Stephen Joseph has pioneered a whole field of mineral modifications to biochar to enhance its properties for applications like filtration but also for use as super-charged biochar fertilizers. Dr. Joseph was inspired to investigate this avenue during the early years of the biochar industry because biochar was hard to come by, and he wanted to maximize the impact of the biochar that was available to farmers.[2]

Examining the black carbon particles extracted from terra preta soils, Stephen Joseph found that the surface of a terra preta carbon particle is coated with nanoparticle oxides of manganese and iron that are bonded together with large organic molecules. These will naturally develop over time as biochar reacts with minerals in soil, but again, he wanted to find out if there is way to accelerate this natural aging process. After years of experiments, he can now formulate composites to enhance development of organic macromolecules and other fertilizer properties.[3]

There is a lot to learn about how to combine biochar with minerals, organics, soil types, and crops, and it's fair to say that the biochar research community has barely gotten started, even with all the recent advances of Dr. Joseph and others.[4] There is plenty of room for small farmers to experiment with making biochar mineral complexes using the material that they have on hand.

It is also interesting to note that cooking biomass and minerals together has been a traditional method of improving soil fertility, as clay was either mixed with crop waste or piled on top of it to control the charring process. In a nineteenth-century book titled *A Muck Manual for Farmers*, author Samuel L. Dana discusses the availability of heaps of waste wood, peat, turf, and sawdust, and recommends that farmers:[5]

> Make these into small piles with loam, giving enough combustible matter to allow the whole, when covered with sods and fired, to burn slowly like a charcoal pit, a slow, smothered fire, burning without consuming, till the whole soil with which the peat or earth, &c., were mixed, has been baked quite up to a low, red heat, just visible in the night.

Dana then recommends using this mixture of "baked soil and charcoal" for the "speedy conversion" of night soil (humanure) to make a valuable fertilizer. He recommends the addition of copperas (iron sulphate):

> Be it understood, that a few pounds of copperas, dissolved in a pail of water, say half a pound to a gallon, will when thrown into an ordinary vault, while its contents are being removed, completely deodorize that, no noxious gases escape, polluting the air, and making night more terrible than day to the dweller in crowded cities.

Furthermore, he says:

> The addition of copperas-water actually adds a direct value to the night-soil. The ammonia so volatile, the sulphuretted compounds so horrible, are seized by the copperas water, they are retained long enough to be transported . . . to the pit prepared for its reception; and then mixed . . . with the charred material.

The farmer will be motivated to undertake this labor because: "Both the ammonia and the sulphuretted hydrogen are too valuable to be lost. No grain ever grew that did not contain sulphur compounds."

The role of biochar and biochar mineral complexes in converting waste to value and completing the circular economy deserves far more attention than it gets.

Water Filtration

Bone char has a long history of use not only as a fertilizer, but also in water filtration. The composition of charred cattle bones was analyzed as 8 to 11 percent carbon, 6 to 9 percent calcium carbonate, 16 percent soluble phosphate and 70 to 76 percent tricalcium phosphate, with trace amounts of calcium sulfate, soluble phosphate, and iron.[6] It also had a specific surface area of 80 to 120 square meters per gram.

To put that in context, the biochar we make from wood in flame-cap kilns and TLUDs usually has specific surface area of around 300 square meters per gram, while common activated carbon used in water filters has 800 to 1,200 square meters. The specific surface areas found in soil constituents can range from less than one tenth of a square meter per gram for sand, to clay with up to 800 square meters per gram.[7]

Surface area refers to all of the surfaces surrounding the tiny pores in a material, and it is the relatively high surface area combined with the calcium and phosphorus that make bone char an effective filtration medium for polluting metals like fluoride, cadmium, chromium, and uranium. Interestingly, bone char can be more effective at removing these metals than more expensive activated carbon with more surface area. That is because calcium and phosphorus integrated with the char surfaces can bond with metal pollutants through several mechanisms, including ionic bonding, hydrogen bonding, and electrostatic attraction. The sorption and removal of contaminants depends on both the porosity (the "sponge") and the electrochemistry of the minerals and organic molecules that are present on those surfaces—the "electric" part of the "electric carbon sponge" described in chapter 2.

Bone char, often called animal charcoal, was used in the past to remove the brown color from cane sugar syrup to make white sugar, and to remove fabric dyes from water. Bone char is still used today to remove fluoride and uranium from drinking water. A decade ago, a group of students and researchers from Oglala Lakota College and the University of Illinois completed the initial phases of a project to evaluate the use of bone char to clean water from private wells on the Pine Ridge Reservation in South Dakota. Nearly half the wells sampled had unsafe amounts of contamination with arsenic and uranium. They were interested in bone char as an indigenous and affordable material compared to very expensive activated carbons that are difficult for the residents to afford. The bone char was effective in removing uranium, but not so great at arsenic removal. Arsenic comes in many forms, so adsorbents often have to be modified for specific wastewater conditions like pH and oxygen content.

Biochar researchers have proposed a multitude of ways to enhance biochar for different uses, including filtration of specific contaminants, but these engineered materials are energy intensive and expensive to produce.[8] Bone char can work well as is, and it can be affordable in places where there is no other alternative to effectively to filter and clean water.

Dr. Josh Kearns is an environmental engineer and the director of science for Aqueous Solutions in Thailand as well as the chief technical advisor for Caminos de Agua in Mexico. Kearns has pioneered the use of locally produced wood-based biochar as an affordable water treatment to remove organic chemical toxins such as herbicides and pharmaceuticals, and bone char for removal of inorganic chemical toxins such as excessive fluoride. Kearns is the author of A Field Guide to Biochar Water Treatment that is published on his Substack site as a living document of discoveries and practical techniques for low-tech water treatment.[9]

Alkaline groundwater can mobilize fluoride from rocks where it leaches into wells. A global assessment of fluoride contamination of groundwater revealed hot spots in central Australia, western North America, eastern Brazil, and widely distributed smaller areas in Asia and Africa.[10] Bone char is effective in treating this groundwater because the fluoride ions are attracted to ion exchange sites on the bone char surfaces. Additional complexation reactions may also occur that keep the fluoride ions attached to surfaces inside the pores of the bone char.

Here is how Josh Kearns made bone char for his water filtration systems:

> In brief, cow bones were cut into pieces, sun dried for several days, and loaded into a 95 L (25 gal) lidded steel drum retort. The retort was heated to reach an internal temperature of 570°C over a two-hour period (4.5°C/min heating rate), held at approximately 570°C for 45 min, then cooled rapidly with water to handling temperature.[11]

Once the bone char is cooled, it is then ground with a mortar and pestle and added in a layer to the filtration column that consists of several other layers of pre-filter materials such as gravel and sand.

One of Kearns's discoveries was that biochar made from wood waste in a drum TLUD like the Big Whirly is the most effective biochar for water treatment filtration. He compared biochar made in an oxygen-starved retort kiln with TLUD gasifier biochar that is exposed to oxygen during production. Retort biochars that had the same amount of surface area as the gasifier biochar did not perform as well in filtration. He surmised that the reason was the "high sweep-rate of air through the reactor that (1) allows for slight combustion of the newly formed char, driving temperatures higher, and (2) vigorously sweeps pyrolysis gases upwards and away from the hot char bed."[12] The effect is to "ream out" the char's pore structure as the oxygen molecules attack the residual chars and oils in the pores of the char. In batch tests of the sorption of the herbicide compound 2,4-D, Kearns found that on a surface area–normalized basis, high temperature (at least 850°C) gasifier biochars exhibit equivalent 2,4-D adsorption performance to activated carbons.[13]

Kearns learned another trick to make gasifier biochar even more like activated carbon—quench it with water instead of snuffing it. He first quenched his biochar batches by plugging all the holes in the bottom of the TLUD and clamping a lid over the top to exclude air while it cooled. That made good biochar, but when he switched to dumping water into the drum to quench it, he found that it changed the properties of the biochar such that the adsorption of the 2,4-D herbicide increased by 40 percent. He hypothesized that the water turning to steam in the process worked somewhat like the industrial method of using high-pressure steam to activate charcoal by opening up pores and increasing surface area.

Experimenting with mineral modifications to biochar, Kearns's colleague Dr. Matt Bentley soaked pine sawdust in ash solutions prior to charring it. He found that it increased the biochar's ability to sorb organic micropollutants by more than ten times. If you are wondering where to get the ash solution for soaking your biomass feedstock prior to charring it—that's easy. It is just the potash lye used in soapmaking that people have used for centuries. Simply take a bucket of wood ash and dump it on top of a few inches of straw laid over some pebbles in a barrel with a slow drain. Pour water through the layers to leach out the highly alkaline lye. You might need to pour the same water through several times. The lye water will have high pH of 10 to 12, consisting mostly of potassium hydroxide and containing other alkaline metals such as sodium, calcium, magnesium, and anions of silica and phosphorus. When the wood soaked in lye is then charred, all these chemicals become embedded in the surfaces of the biochar where they use their electrical and molecular bonding powers to snag the nasty chemicals you do not want to drink.

Using iron to modify biochar surfaces is another technique that can improve biochar's ability to capture anions like phosphate, nitrate, and negatively charged mineral pollutants like arsenic.[14] Theoretically, there are several low-tech ways of adding iron to biochar: You could soak biochar feedstocks in iron sulfate or rusty water and dry before pyrolyzing; you could sprinkle an iron-bearing clay like bentonite in the kiln while you are pyrolyzing; or you could quench biochar with an iron solution. These techniques need to be tried and tested.

By the way, you can make iron sulfate by soaking iron pyrite (fool's gold) in water. Ian Donaldson describes the process on the website Julian's Science Experiments.[15] Donaldson puts a layer of iron pyrite nodules in a plastic bucket and adds just enough rainwater to cover it. Using rainwater is important because chlorinated water can kill the bacteria responsible for the process. The bucket stands in a warm place for several weeks until the water evaporates and the black nodules are covered with a green and white crust. The crust is formed by extremophile organisms that eat the iron pyrite and deposit iron sulfate as a waste product.

This is the stuff known as copperas that was used to deodorize the night soil as discussed in *A Muck Manual for Farmers*. Adding some copperas to your biochar may enhance its ability to remove anions like nitrate, whether you are filtering water or using it in the outhouse.

Biodegradation

Another advantage of biochar as a filtration medium is its friendly relations with bacteria. The cleansing power of biochar is enhanced by the microbes that adhere to its surfaces. Formation of a biofilm in the water filter is unavoidable. While a biofilm adds more organic material that can foul the filter, a healthy biofilm will destroy pathogenic bacteria like *E. coli*. The biofilm works synergistically with the biochar to degrade contaminants in a process called bioregeneration. In bioregeneration, the combined adsorption-biodegradation process is more efficient than either the adsorption or biodegradation processes alone. According to Josh Kearns, the biochar adsorbs the dissolved contaminants allowing more time for their breakdown by the biofilm, which in turn frees up surface sites on the carbon for additional sorption, extending the life of the filter media. Even some compounds typically classified as non-biodegradable are broken down in long-running carbon bio-filters. Exposure to contaminants retained by the carbon over periods of weeks to months allows microorganisms to acclimate and develop the enzymatic pathways necessary to break down some otherwise resistant compounds.

In other words, the biochar holds down the bad guys and the microbes keep pummeling them until they are turned to dust! This biochar-biofilm partnership is also used very effectively to clean ponds of nitrates and phosphorus that cause algal blooms. I helped a friend add several buckets of biochar to his koi pond. The green algae were so thick that the colors of the koi fish were muted to brown if you could see them at all. Now you can sit on the bench beside the pond and watch the beautiful orange and red koi as they swim in the clear water. On a larger scale, many cities that have algae-filled ponds and lakes are now using biochar "socks"—filter bags filled with biochar—to absorb the nutrients and clean the water with great success. Dairies are also using biochar in their wastewater lagoons.

Cities and towns can keep those nutrients and other contaminants out of water bodies in the first place if they capture storm water runoff in bioretention basins with biochar. These are swales that are filled with compost and sand to filter out metals, nutrients, and pathogens from stormwater before it goes into a river or a water treatment plant. Biochar is able to not only extend the life of the bioretention filters but improve their efficacy. One research project added fungi and biochar to the standard mix of compost and sand. The combo worked well to remove PAH chemicals and fecal coliform bacteria.[16]

For an excellent guideline to many different applications of biochar in stormwater management, see the Minnesota Stormwater Manual found online at stormwater.pca.state.mn.us. Potential applications of biochar include downspout filter boxes, tree planter boxes, green roofs, filtration socks, constructed wetlands and ponds, sand filters, swales, and infiltration basins.

Clean Water for All

Josh Kearns has detailed instructions for making a water treatment system that includes gravel roughing filtration, a slow sand biofilter, and a charcoal finishing filter. Figure 7.3 shows a point-of-use charcoal filter that can be set up in a kitchen for the final filtration step before use.

The work that Josh Kearns is doing has wider implications than the remote village communities where he has spent years developing affordable filtration systems. In an article published in *Resilience* magazine, he explained:

> Development professionals do their work under the assumption that the developing world will someday look a lot like the developed world. But there's a good chance that they're wrong. A practical look at the world's energy supply, and interesting new research into the link between energy, culture and quality of life, shows that the reverse is probably true: The developed world will soon look more like the developing world.

Figure 7.3. A biochar water filter is demonstrated at a workshop for developing methods to address petrochemical pollution in the Ecuadorian Amazon. Photo courtesy of Josh Kearns

> Here's why that's happening and what we can do to prepare for a big change right now.[17]

Even today in the United States, many people do not have access to truly clean water. Native Americans whose reservations are downstream from uranium mines are not able to access funding to clean their water.[18] The predicament of Flint, Michigan, with its lead-contaminated water supply is well known. Nationwide, drinking water treatment infrastructure is deteriorating and the chemical burden of per- and polyfluoroalkyl substances (PFAS) and other persistent organic pollutants is growing.

Being resilient and self-reliant depends on access to clean water, so it's great news that people like Josh Kearns are working on these methods. Water filtration is very complex and finding better ways to do it on the cheap is vital work. I recommend that you consider supporting Josh Kearns by subscribing to his newsletter at JoshKearns.substack.com where you can get all of his updates on affordable techniques for filtering and providing clean water.

CHAPTER EIGHT

A Modern Muck Manual

Every animal makes waste or "muck." We all eat and excrete, and perhaps that is the very essence of our existence. Humans are no different from any other animal that way, except that no other animal spends as much time as we do managing our own muck. With the exception of ants, prairie dogs, and other colony-forming creatures, no animal that I know of spends any time at all managing its muck; they make their droppings and move on. Poor us, we not only have to manage our own muck but also that of our livestock and our pets.

Modern methods of muck management leave a lot to be desired, as nearly all the fertilizing power of human sewage and animal manure is wasted. As long as we have fossil fuels to make our fertilizers, we can afford to flush it all away, but when the day comes that we cannot afford to waste these nutrients, we had better have another plan in place. I have started collecting some of the many "muck manuals" that were published in the nineteenth century. Thousands of pages were devoted to describing all of the different ways that manure could be managed and treated to improve its effectiveness as a fertilizer. Many writers included charcoal as a treatment. For these writers and the farmers who read their words, muck was essential to life.

Today we have access to a lot more scientific knowledge about the chemical and biological processes that take place in composts and ferments. We also have a broader knowledge of practices outside of Western economies that we can hope to master and combine into a modern compendium of knowledge about "muck."

Biochar can solve two major problems with the current use of manure as fertilizer: nutrient loss and pathogens. By preserving nitrogen, biochar preserves the correct ratio of nitrogen to phosphorus. Because manures are so badly managed, they lose half or more of their nitrogen before they get to the field. As a result, high-phosphorus manures, when applied at appropriate rates to provide the necessary nitrogen to a crop, will add excessive amounts of phosphorus to soil where it can leach into groundwater. Adding biochar to animal manures at livestock facilities will produce a higher-quality compost with balanced nitrogen and phosphorus, ready to apply as fertilizer to fields. Adding biochar will also help control pathogens like *E. coli* bacteria.

Biochar in Bedding

Biochar has a considerable capacity for sorbing ammonia, but other factors, particularly pH, must be controlled for best results. The process of sorption has different mechanisms and can be impacted by temperature, pore volume and size, available surface area, and the chemistry of the surface area. Biochars that have acidic functional groups on their surfaces will do better at adsorbing the alkaline ammonia gas.

Biochar applied directly to bedding can reduce ammonia emissions, especially if it has high surface area and it has been treated to lower the pH. The addition of non-acidified peanut hull char to poultry bedding did not reduce ammonia concentrations in the air compared with untreated pine shavings bedding material. However, the use of acidified chars resulted in significant reductions (about 50 percent) in ammonia concentrations.[1] Another option is to use wood vinegar as the acidifier. Pig manure composted with biochar and as little as 1 percent wood vinegar decreased emissions of ammonia, methane, and other greenhouse gases.[2]

The environment in a commercial poultry operation is extreme. A poultry house is typically cleaned after each six-week cycle of raising chicks to broilers, and workers must wear respirators before entering the houses to protect against inhaling ammonia. Ammonia levels in poultry barns are generally controlled by using acidifiers in the bedding. Acidifiers prevent action of the urease enzyme that mineralizes urea into ammonium. Common acidifying litter treatments include citric acid and sodium bisulphate. Citric acid is longer lasting but more expensive and mainly used in organic chicken production. One problem with relying on sodium bisulphate alone is that once the chemical has reacted with the ammonia, it becomes neutralized and is no longer effective. Generally, this neutralization occurs before the grow cycle is complete. A typical practice is to add just enough acidifier to prevent young chicks from croaking during their first three weeks of life. After that, operators let the ammonia levels rise rather than reapply expensive acidifiers.

Researchers conducted experiments that compared acidified biochar to sodium bisulfate as an acidifier in poultry barns. The acidified biochar worked as well as sodium bisulfate but had to be applied at twice the rate. With an application rate of 200 pounds (91 kg) of acidified biochar per 1,000 square feet (93 m^2) of poultry house, ammonia was reduced up to 47 percent.[3]

One approach that has great promise is to apply biochar along with an inoculant that includes lactic acid bacteria and molasses. As the lactic acid bacteria work to break down the litter, they produce organic acids that acidify the biochar.

Another aspect of nitrogen preservation is the incorporation of nitrogen in proteins made by microbes. To get this happy result, microbes need to have access to degradable carbon that they can combine with the nitrogen to grow and multiply, just like compost needs the proper carbon to nitrogen (C:N) ratio to really take off and heat up. In other words, if there is plenty of carbon available to balance the nitrogen, the microbes can eat the ammonia and multiply, turning it into microbe biomass.

Research has confirmed the theory that degradable carbon helps to immobilize ammonia nitrogen by increasing microbial activity, which ties up nitrogen in organic forms (dead microbes) in the microbial biomass. The solution can be as simple as adding a highly degradable carbon source such as molasses to the manure. One study found that when biochar is used in conjunction with another highly recalcitrant bulking material like sawdust, it may be necessary to supplement with molasses or some other highly degradable carbon.[4] Another study added biochar to compost with poultry litter and apple pomace, a source of highly degradable carbon. In this compost mixture, with abundant levels of both nitrogen and degradable carbon, the biochar improved composting conditions and accelerated the compost process.[5]

Researchers at the University of Hawaii have developed a successful biochar inoculated system for piggeries, called the Inoculated Deep Litter System (IDLS).[6] The IDLS starts with a base of 6 inches (15 cm) of biochar mixed with cinders, topped with 2 feet (61 cm) of logs and a foot (30 cm) of green waste chips, all inoculated with *Lactobacillus* and other indigenous microorganisms from the local environment. The IDLS is not removed for cleaning, because it breaks down and composts in place. It is not aerated, maintaining a condensed pack of manure and organics. Periodically, new wood chips are added. Once the proper microbial environment is established, the bedding composts on its own, with no objectionable odors, and the animals are healthier as a result.

Biochar added to dairy bedding can help prevent environmental mastitis and other cattle diseases by slowing the spread of pathogens. Researchers in Switzerland attribute a large part of the success in improved dairy conditions to adding 10 percent biochar to bedding material.[7] They found that biochar sorbs moisture in bedding and inhibits ammonia emissions. Reduced odor in dairy facilities benefits both cows and workers.

Biochar First!

We do well to remember that the making of muck begins in the gut. Eating precedes excreting, and for the best, most effective muck management, that is where we should start.

In the spirit of first things and going first, I will describe my own intake of biochar. I clean my mouth with biochar every day, using my own recipe for charcoal toothpowder. This routine is healing my gums. I used to dread going to the dentist, having the hygienist tell me that my gums had continued to erode as my "pockets" (the measured depth of erosion) grew deeper every year. About five years ago I started using charcoal toothpowder exclusively and a couple of years after that, the hygienist was astounded to tell me the pockets had begun to shrink. "That doesn't happen," she said, and asked me what I had been doing. I told her about the toothpowder and she was surprised. She said she had other patients who used charcoal toothpaste but did not get the same results. My conclusion is that the powder is better than the paste, and certainly cheaper.

A 4.7-ounce tube of charcoal toothpaste from a major brand runs about $15 and may last you several months. I can buy all the ingredients for about a gallon of toothpowder for less than $60, and that includes hardwood charcoal powder and mint powder for flavoring. If I use my own charcoal and powdered dried mint, it would cost me about $30 to make a gallon of toothpowder. I have not monitored our use, but I would say that our two-person household goes through about 12 ounces of toothpowder a year, so it would take us a decade to use the whole gallon. That's why I packaged most of it up into 2-ounce jars to give as gifts. Here is my recipe, in case you would like to make it and maybe turn it into a business selling charcoal toothpowder at your local farmers market. Do be aware that xylitol is toxic to dogs, so omit this if you are making a canine toothpowder. This is for one small jar, but you can scale it up for big batches:

Charcoal Toothpowder Recipe

2 tablespoons (30 ml) food-grade hardwood or coconut powdered charcoal
1 tablespoon (15 ml) food-grade bentonite clay
2 teaspoons (10 ml) baking soda
2 teaspoons (10 ml) xylitol sweetener (optional)
2 teaspoons (10 ml) peppermint powder or cinnamon

Gut health starts in the mouth, and using charcoal toothpowder controls bad bacteria. I also eat a lot of fermented foods that I make, like sauerkraut and pickles. When my digestion feels off, I take a spoonful of powdered charcoal mixed with 12 ounces (355 ml) of water to make a charcoal slurry and it rights things pretty quick. I take mineral supplements and eat nutrient-dense food from my vegetable garden. In general, I drink a glass of charcoal slurry about every other day.

Then I make muck. I can certainly tell when I have had charcoal because it shows up there. I have used various systems for muck management, including an outhouse for the thirty years that I lived in an off-grid cabin. Now I have a flush toilet and a septic system, but use a bucket setup out in my shop. The bucket system is very successful because in addition to biochar, I use bokashi.

Bokashi EM for Effective Muck Management

Bokashi is a Japanese term for fermented fertilizers. It also means "obscured effect," which is appropriate because the impact of bokashi is subtle but profound. Basic bokashi is made with wheat or rice bran using a liquid culture called EM-1 or Effective Microorganisms. EM-1 is a consortium of microbes that all work together. This culture consists mainly of lactic acid bacteria (LAB), yeast, actinomycetes (a soil bacteria), and purple non-sulfur photosynthetic bacteria (PNSB). EM-1 was developed and commercialized by a Japanese researcher, Dr. Teruo Higa, and has been in use throughout the world for decades as a soil amendment, composting aid, water treatment, and health tonic.

The PNSB is interesting because it plays an important role in keeping the microbial community healthy. According to Higa and his coauthor James F. Parr of the USDA Agricultural Research Service, PNSB can transform a putrefying anaerobic process into a healthy fermentation by consuming hydrogen sulfides and other bad-smelling compounds, using sunlight to drive the process.[8] This certainly has relevance to processing our muck! Author Adam Footer, in his book *Bokashi Composting*, has a good summary explanation of the power and adaptability of PNSB, as well as many other tips and tricks for making and using bokashi.[9]

While you can find instructions on how to make your own effective microbe cultures from forest soil incubated on cooked rice, I find it easiest to buy a bottle of the EM-1 culture and propagate it in a simple process like brewing beer or kombucha. See chapter 11 for propagation instructions.

Terra Preta Sanitation

Dr. Ralf Otterpohl of the Hamburg University of Technology and his team have developed and promoted the concept of terra preta sanitation, a sanitation system that uses biochar and *Lactobacillus* or bokashi in the context of households and villages, and perhaps even towns and cities. The concept is modeled on the apparent practices of the ancient Amazonians who built the fertility of their soils with charcoal and human wastes.

We have embraced the Amazonian practice of adding charcoal to soil. When will we also adopt the practice of gifting our own excreta to the soil microorganisms? Can we call ourselves an advanced civilization when we still engage in barbaric practices like pooping in our own drinking water? Even if you are stuck with a flusher, it is good to be prepared for disaster in the event that water or power outages make the flusher unusable. Terra preta sanitation is good as either a primary or a backup sanitation system. The system rests on the following principles.

Separate collection of urine and feces as much as possible. Provide charcoal and microbial inoculant to households to add to the collection containers. Store the containers for four weeks and then collect the material with trucks and transport to a service site where urine is composted with green waste, and feces are vermicomposted. Lactic acid fermentation for four weeks kills most bacterial pathogens because microbial pathogens cannot survive the acidic conditions. Six months of vermicomposting should take care of the rest. Vermicompost also kills any parasitic worm eggs that might be present.[10]

Lactic acid fermentation is what produces yummy foods like sauerkraut, yogurt, and Asian pickles. Pickled poo may not be "yummy" but it has one great virtue—it eliminates the poo smell. That's right, pickled poo does not stink. It doesn't smell wonderful, but it's not really offensive. You can keep your bucket indoors right in your bathroom and no one will complain.

Here's what my terra preta sanitation system consists of:

- Two buckets—for pee and poo—with spin-on lids. You want these because the buckets need to be sealed and regular bucket lids are hard to pry off—especially when you are in a hurry. You can also buy a bucket with a toilet seat and lid called a Luggable Loo. I have a couple of these as well.
- Bokashi.
- A supply of biochar, crushed to half an inch (1 cm) or smaller.
- A place for processing the poo. The poo goes into a large barrel for storage and then is applied to trees. I could use it in the garden but feeling cautious, I have only used it on trees so far.
- A place for processing the pee. The pee-soaked char goes in the compost or sometimes directly in a garden bed. I have also used it in potting mix as a fertilizer. It is amazing!

To use my system, I start by adding about 4 inches (10 cm) of biochar in the bottom of the poo bucket and sprinkling a cup of bokashi on top of that. I fill the pee bucket about three-quarters full with biochar and sprinkle two cups of bokashi evenly across the top of the char. Then I fill a container with a biochar/

bokashi mix to set beside the buckets, along with the usual roll of toilet paper. Here are use instructions posted on the inside of the bathroom door:

For Poo

Make your deposit.
Drop the toilet paper in the bucket.
Sprinkle a handful of bokashi on the deposit.
Close the lid.

For Pee

Make your contribution, but it needs to be urine only, no poo. It is OK to pee in the poo bucket, but not the reverse.
If you use toilet paper, put it in the trash can, not the bucket.
Close the lid.

Terra preta sanitation is not just for the more intrepid among us. It could also become a mainstay in disaster relief, helping those forced into situations out of their control. In 2020, a fearsome, hot Santa Ana wind started blowing west across the Cascades over the Labor Day weekend. Multiple ignitions grew into a firestorm that wiped out thousands of acres of wild forests and choked all of the West Coast with heavy smoke for weeks. My community was evacuated and many had nowhere to go. My friend Brenda Patton, who sponsors our biochar kids' camps, invited everyone in our neighborhood to shelter at the Enchanted Forest, her 250-acre rural property, out of the way of the flames.

Brenda had everything she needed to care for a group of fifty people, except for one thing: porta-potties. There were none to be had from all the local rental places, and her aging septic system could not handle the deluge. She called me and we came up with a plan to use biochar and bokashi in improvised outhouses with buckets. Terra preta sanitation saved the day, taking care of our community and keeping them safe from flies, odors, and pathogens while they were away from their homes. Luckily, while there were close calls, no homes were lost in our community.

If you are not quite ready to compost poo, a biochar urinal is a good place to start. Nutrients in urine are significant. According to composting guru Sally Brown, one person will produce about 6.6 kilograms (14.5 lbs) of nitrogen and .91 kilograms (2 lbs) of phosphorus during a year's time. She says, "Those numbers are remarkably close to the amount of N (6.6 kg) and P (1.1 kg) we need each year to grow food. And it all goes down the sewer."[11]

Unless someone is very ill, their urine is mostly sterile. The challenge with urine is to prevent the formation of ammonia from urea, and the loss of nitrogen as the ammonia volatilizes. This is where bokashi becomes essential.

Figure 8.1. Two types of biochar urinals. The one with the screw on lid is more secure (*right*), but the one with the seat is more comfortable to sit on (*left*). No worries about splashes—biochar absorbs liquid very easily. You can see the bokashi grains sprinkled on the surface.

Alkaline pH facilitates the ammonia volatilization. We want the ammonia to transform to nitrate and be absorbed by the biochar. Adding a thin layer of bokashi across the top of the biochar in the bucket seems to work very well. Without the bokashi, the bucket can smell like ammonia. With the bokashi, there is no smell. This is really worth doing because urine is the fastest, easiest, cheapest way to pre-charge your biochar with nutrients. Plus, it is a shame to waste all that free fertilizer that you and your family can generate.

A few things to consider when using urine as a fertilizer:

- Pharmaceuticals may remain in the urine. A period of composting with biochar may help break down those chemicals, but we need more information.
- Urine is generally sterile, but don't use urine from people who are really sick with infections.
- Check out the book *Liquid Gold* by Carol Steinfeld for more information.[12]

What is good for humanure is good for livestock manure. The combination of biochar and bokashi or *Lactobacillus* seems to be the answer to just about every manure management problem. Many small farmers and gardeners have begun using biochar in chicken coops, goat barns, pigpens, and horse stalls, reporting that smells are almost completely eliminated, and animals are healthier.

From the Forest to the Barn

In 2015, the South Umpqua Rural Community Partnership (SURCP) was awarded a grant from the Natural Resources Conservation Service (NRCS), a division of the USDA: "On-Farm Production and Use of Biochar for Composting with Manure." SURCP has a mission to conserve the forests and rivers and promote sustainable living in their watershed. Stanley Petrowski, with Jim and Barbara Long, started the Umpqua Biochar Education Team when they realized that biochar could help everyone in their community. SURCP requested the funds to work with small farms in southern Oregon that had both forestland and livestock. Forest fuel load reduction, forest health treatments, and logging operations produce abundant debris on forestlands in Oregon that could be used to make biochar. Customarily, the debris is burned to ashes in the open air, reducing the fuel load danger, but producing considerable smoke pollution. SURCP hired me to manage the project.

Our vision was to see if we could combine two farm waste streams—woody debris and animal manure—in order to make valuable composts and organic fertilizers for use on the farm. We spent about half of our time developing kilns (the Oregon kiln and the Ring of Fire kiln) and making biochar, and the other half using the biochar to manage manure.

Because we had a variety of producers using different systems for composting and manure management, we experimented with a lot of different processes. The most efficient process is to use biochar directly in the barn where it can capture nitrogen from urine and manure as it is generated. However, some barns are cleaned daily and it is more convenient to add biochar as barn scrapings are piled for composting. Farmer participants in this project used several different manure management systems. These can be summarized as:

Pack barns, where manure and bedding build up over several months until they are cleaned out and piled all at once. Some barns use very little bedding, and others use a lot. Some piles are mixed with wood chips for better composting. Some piles are not mixed with anything else, just piled as pure manure.

Barns with daily or frequent cleaning where manure is cleaned out and added to a compost pile. There is very little bedding in these systems and no extra carbon is added to the compost piles.

Farms without barns. Some producers had no animals on the farm and they imported manure from elsewhere in order to build hot, thermophilic compost piles from other farm waste materials.

Our goal was to design a protocol for each producer for using biochar in their preferred manure-management system to the best effect. To help guide our efforts, we began by investigating how others have approached the problem, with the following findings.

A manure pack in the barn can be managed or unmanaged. An unmanaged pack is simply the accumulation of manure and bedding in the barn that is cleaned out at regular intervals. A manure pack may also be managed to accelerate composting in the barn. Typically, this is done by a combination of acidifiers to prevent ammonia volatilization and regular rototilling of the pack to create more aerobic conditions for composting.

To avoid the work of rototilling and the expense of acidifiers, we looked toward practices developed in Asia that use *Lactobacillus* to acidify and charcoal to aerate, absorb nitrogen, and improve the composting process, like the inoculated deep litter system for pig manure compost discussed previously. Here are a few examples of the farms we worked with and what we learned. Hopefully these results can help others do their own experiments using biochar and microbes to better manage manure.

WILLOW WITT RANCH

Willow Witt Ranch has a small goat dairy barn with a manure pack that builds up over several months between cleanings. Along with some straw bedding, goats drop some of their alfalfa feed onto the floor as they eat. As a result of goats' messy eating habits, a lot of alfalfa is wasted and becomes bedding. Willow Witt recently started a new program of feed control to improve animal health by restricting feed. As a result, the animals don't waste as much and there is less bedding. Willow Witt was faced with either having to purchase straw for bedding, or start using biochar.

Figure 8.2. Biochar packed into 55-gallon drums is fermented with bokashi (made from wheat bran, EM-1, and molasses) for several weeks, lowering its pH from 9.5 to about 6. The white growth on top is the beneficial actinomycetes filamentous bacteria.

We tried two methods at Willow Witt. The first year, we took the biochar made on the ranch and packed it into 55-gallon drums along with bokashi and molasses to ferment for several weeks. This effectively lowered the pH of the biochar to less than 6. The biochar was sprinkled once a week on top of the bedding in the barn. This was successful, but the supply of biochar-bokashi ran out. During the second year, the ranch used plain biochar sprinkled in the barn once a week, followed by spraying an EM-1 solution of 4 ounces (120 ml) of EM-1 and 4 ounces of molasses to 1 gallon (3.8 L) of water. This combination was very successful in completely eliminating ammonia odors. After six months of adding biochar and EM-1 spray in the goat barn, ammonia odor was gone, even when digging into the pack.

Willow Witt Ranch owner Suzanne Willow said, "We were very impressed by the odor-reducing power of biochar. It sure has improved our barns. When you dig into the floor, it looks like it's composting really well. Instead of the plate of waste hay and alfalfa and pee and poop, it's nice compost."

Here are instructions for the two methods that they used:

Method #1

1. Make biochar and crush to ¼ inch (6 mm) or smaller. You will need about 40 gallons (150 L) to fill the 55-gallon drum.
2. To lower pH and inoculate with microbes, prepare 10 gallons (38 L) of bokashi.
3. Prepare molasses water by mixing 16 ounces (475 ml) of molasses with 2 gallons (7.5 L) of warm water.
4. Mix all the bokashi, biochar, and molasses water on a tarp.
5. Pack mixture into a 55-gallon drum to ferment for several weeks.
6. Sprinkle one 5-gallon bucket of inoculated biochar per 100 square feet (9.3 m^2) of barn area once a week.

Method #2

1. Make biochar and crush to ¼ inch (6 mm) or smaller.
2. Spread biochar to dry.
3. Sprinkle one 5-gallon bucket of dry biochar per 100 square feet (9.3 m^2) of barn area once a week.
4. Spray a mixture of 4 ounces (120 ml) EM-1 and 4 ounces (120 ml) molasses mixed into 1 gallon (3.8 L) of water per 100 square feet (9.3 m^2) of bedding surface once a week.

MICHAELS RANCH

Michaels Ranch produces grass-finished Emerald Hills meats, with 200 beef cow/calf pairs, 200 yearlings, and 325 ewes on 450 acres of grazing land. About

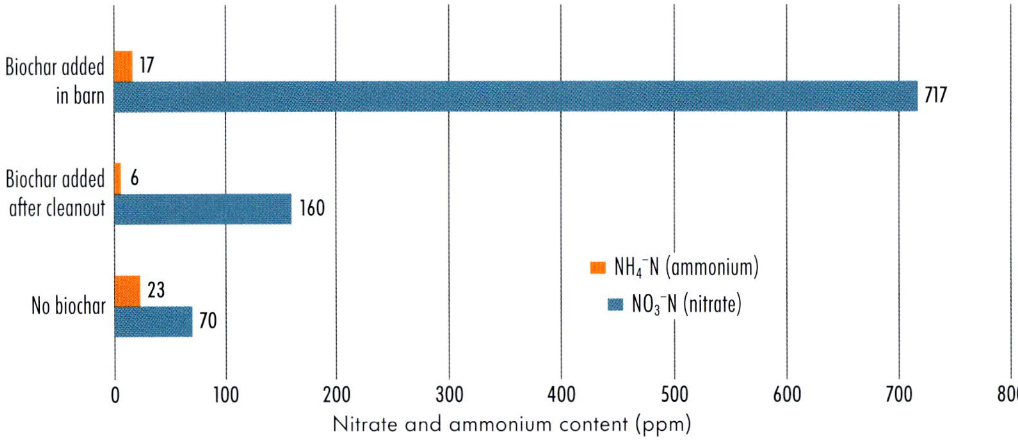

Figure 8.3. Adding biochar to the barn at Michaels Ranch lets the cattle drop manure on it all winter and mix it with their hooves. The result is ten times more plant-available nitrate than in the untreated manure.

120 acres of forest needs occasional management for fuels reduction. Pasture fertility is a constant concern. Troy Michaels typically applies 200 pounds per acre of NPK and S (sulfur) fertilizer on his pastures. He also uses manure from the winter barn where 150 calves are fed every year. The barn is scraped and manure is piled until it can be applied with a manure spreader in different sections of the pasture for fertilizing and increasing soil carbon. He is able to cover about 15 to 20 acres each year with barn manure from two piles of about 30 cubic yards (23 m^3) each. No bedding is used in the barn, so the barn scrapings are almost 100 percent manure. These piles shrink and dry out as the manure breaks down. There is not enough carbon or aeration to produce thermophilic compost.

Odors are not a problem because the barn is very open. The purpose of adding biochar to the winter barn is more for improving the compost than for improving the barn environment. Michaels Ranch uses high-carbon boiler ash (40 percent carbon) as a liming agent on their pastures, so it was easy for them to add some of that material to the barn at the beginning of the winter season, and the cattle mixed it into the manure with their hooves. The mixed biochar-manure was scraped and piled in spring of 2017, and it reached thermophilic temperatures, composting nicely.

The first year we worked with Michaels Ranch, we were not able to get biochar into the winter barn before the cattle entered the barn, so Troy Michaels mixed the manure with biochar in the spring during his normal process of cleaning out and piling the manure. We sent samples for testing of that post-cleanout mixed biochar manure pile (2016), the plain manure from the year before (2015), and the fully mixed-in-the-barn pile from the following year (2017). Figure 8.3 shows the nitrate and ammonium content of the three piles.

Figure 8.4. Michaels Ranch spreads their high-nitrogen, biochar-enhanced manure on spring pasture.

The total nitrate content of the piled manure increased by ten times, while the ammonium content was reduced. Nitrate can leach with rain, but biochar holds on to it so it does not leach as easily. An economic analysis showed that the total cost to the ranch was $200 for the labor of moving and spreading the biochar in the barn. All the other costs were as usual, including applying the composted manure using their manure spreader. This surely seems like a good return of nutrients on a small investment in biochar.

SISKIYOU ALPACA

John and Christine Gardiner make a prized compost from their Siskiyou Alpaca operation that is in demand by local growers. Sales of the "Paca Poo" have helped their bottom line. They agreed to try an experiment with biochar, and we made several compost piles in wire bins with manure that was cleaned out of the barn every couple of days. One pile had 12.5 percent biochar, and a few weeks later we made another pile with 3.6 percent biochar. We made control piles at the same time and kept temperature records for all of the piles. We found that the composting process is very sensitive to the rate of added biochar. Hot composting depends on the starting ratio of carbon to nitrogen in the ingredients, or the balance of "browns" (high-carbon materials) and "greens" (high-nitrogen materials). The best compost piles will have a C:N ratio of 25 or 30 parts carbon to 1 part nitrogen. The question is, what about adding "blacks" to the ingredients? Even though it is almost pure carbon, biochar does not break down to feed microbes, so what does it actually do in the compost?

Figure 8.5. Alpacas are neat and orderly animals. In the barn they all use the same latrine pile, making it easy on the humans who clean up after them.

Figure 8.6. The pile on the left has 12.5 percent biochar added. The one on the right is plain paca poo.

Alpaca manure as it comes from the animal is a pellet that has an almost perfect C:N ratio (25:1) because there is so much undigested fiber in it. Adding a small amount of biochar (3.6 percent) accelerated the heating, but when we added 12.5 percent (3.5 times more), it had a slight cooling effect as compared to the control. Why was the pile with more biochar slightly cooler? Biochar could have changed the ideal C:N ratio, or it could have changed how the pile retains heat. It made little difference in the final compost, as both biochar piles composted beautifully.

We noticed that the consistency of the piles was different: Even after six months of composting, both piles without biochar had a slimy, mucky consistency, while both piles with biochar were much more soil-like with crumb structure. The biochar piles also had a lot of plants and grass growing on top, almost completely covering the piles. The piles without biochar did not have much plant growth on top, indicating that they still had enzymes that would inhibit seed germination.

Figure 8.7. John Gardiner takes a compost sample for analysis. The piles with biochar composted very well and quickly turned into soil where plants could germinate and grow.

Close observation of the samples we took showed that both of the piles with biochar had fewer mature worms than the piles without. Generally, the alpaca manure compost has a very high content of red wriggler worms, so the reduced number of worms was notable, although they were still present. Curiously, the piles with biochar had a very large number of recently hatched baby worms, as compared to the control piles. Did we create a red wriggler nursery? Where did the adults go? Did they die off or migrate out of the pile? The control piles were right next to the biochar piles so they could have headed over there for a better meal. Another biochar mystery to solve!

FRESH CHICKEN MANURE COMPOST

This was an experiment to see how much biochar we could add to manure and still get a benefit. We used fresh chicken manure with very little bedding from a small commercial egg-laying operation and divided it into three piles of 10 gallons (38 L) each. We mixed in biochar at rates of 50 percent, 25 percent, and 0 percent. Despite the small pile size and likely because of the high nitrogen content of the manure, the piles did heat noticeably and we were able to

Table 8.1. Final pH after Biochar Addition to Fresh Chicken Manure and 6 Weeks Composting

	0% Biochar	25% Biochar	50% Biochar
Final pH	8.65	8.25	8

measure some temperature differences between them: Both of the manure piles with biochar were several degrees hotter than the plain manure pile. In contrast to the alpaca manure, which had a more optimally balanced C:N ratio to begin with, the chicken manure was so nitrogen-heavy that we could add up to 50 percent biochar and still improve the conditions for thermophilic composting.

It was interesting to see that the high-pH biochar (close to pH 10) was able to lower the pH of the compost. Also, the final pH of the 50 percent biochar pile was even lower than the pH of the 25 percent biochar pile, even though it had twice as much alkaline biochar added! This must have been due to the power of microbial action and the way that microbes deposit organic acids on the biochar surfaces. We also noticed that the piles with biochar produced a very nice, soil-like substrate with no odor after only six weeks, while the plain manure pile was little changed. Here is yet another example of combining two waste products to get something way more valuable than either of them used alone. This is the miracle of well-managed muck.

CHAPTER NINE

Biochar in Food and Medicine

Manure management with biochar is a win-win because we can take two problem waste streams and combine them to make valuable fertilizers and soil amendments without requiring much in the way of inputs from off the farm. We found that applying biochar right in the barn to catch the manure as it comes out is the most efficient and effective way to use it. But what if we could have started at the other end and added it to the animal's feed?

Australian farmer Doug Pow has experienced profound effects from feeding biochar to his pasture-grazed cattle. The animals are healthier but the biggest change he sees is in his pasture, where nature is taking over and doing his work for him. As soon as a cow makes a dropping, the dung beetles go to work on it, burying manure and biochar deep in the ground. Pow sees more grass and earthworms, and legumes like clover sprouting after the dung beetles work. Cows on clover do so well that he has stopped buying any supplemental hay, and their health has improved to the point where he no longer needs pesticides to kill parasites. One small change, feeding biochar to his 60 cows at a total cost of about $1,000 a year, saved him $20,000 in costs for hay and pesticides, while at the same time, his pasture land is being permanently improved with biochar without having to add fertilizer or use tractors to till the ground.[1]

These results should interest any cattle farmer. There has also been a lot of interest recently in using biochar in cattle feed because it could help to reduce methane eructations (burps) from cattle that occur when they are not processing feed efficiently.

The digestive system of a ruminant animal is rather like a farm. The first part of its multipart stomach, the rumen, is full of microbes that break down the high-cellulose forage that most cattle eat. This is like a compost pile or a fermentation vessel. The cow does not absorb the food at this stage. Then the material passes through to the rest of the gut and now is full of microbes that have converted the cellulose sugars to protein. This "compost" is what the animal absorbs as food and it includes the dead microbes themselves, which are 40 to 50 percent protein. This is why cattlemen often say that they "feed the bugs, not the cow," just like farmers say "feed the soil, not the plant."

Methane

Methane is produced in the rumen because microorganisms that digest fiber produce hydrogen (H_2) gas as a waste product. H_2 gas combines with carbon as the terminal electron acceptor in the reaction, producing methane, CH_4. This reaction represents an energy loss to the animal because the CH_4 is purely a waste product that must be eliminated through eructation. Many attempts have been made to find alternative electron acceptors to lower CH_4 production, but the only alternatives in the anaerobic environment are nitrogen or sulphur, producing ammonia or hydrogen sulfide, both of which are toxic to animals at certain levels.

Some feed supplements can alter the microbial balance in the rumen, selecting for microbes that produce less H_2 in their metabolism and make more efficient use of food energy. One widely used supplement that both reduces CH_4 and increases weight gain is a class of antibiotics called ionophores. Ionophores (at a newly prescribed higher dosage) were identified by the government of Alberta, Canada, as a feed technology that directly reduces enteric CH_4 emissions.[2]

Ionophores work by inhibiting the bacteria that produce methane—the methanogens. Other chemical feed additives that work this way are 3-nitrooxypropanol and various bromine compounds (also found in seaweed). These pharmaceutical and chemical additives all have potential long-term side effects such as excess hydrogen in the rumen, and they all tend to lose effectiveness as the methanogens adapt and evolve resistance to them.[3]

Supplementation with biochar could become a viable alternative for achieving the dual objectives of reduced methane emissions and increased weight gain. A series of studies by scientists working at Souphanouvong University in Laos looked at the effects of different biochars on enteric emissions and weight gain in tropical cattle.

They began with the hypothesis that biochar promotes growth of methane-gobbling methanotrophic bacteria in anaerobic systems. The hypothesis is based on the observation that biochar applied to rice paddies reduced methane emissions by providing habitat sites for methanotrophs that consumed the methane gas. They speculated that a similar phenomenon could occur in the rumen, as previous studies had isolated methanotrophs living in the rumen fluid. They further speculated that two factors normally limit the number and effectiveness of methanotrophs in the rumen:

1. Methanotrophs grow slowly in anaerobic environments where they are limited by the energy availability. As it moves through the animal's digestive system, rumen fluid has a short turnover time that prohibits substantial growth of methanotrophs. If biochar in the rumen provides a favorable habitat for methanotrophs, it could support their proliferation.
2. Methanotrophs are normally attached to rumen epithelium, but to be effective in capturing methane, they would have to be spatially distributed close to the site of methane production on the biofilms that are closely attached to feed particles. Biochar might provide the means to bring them in closer proximity.

These researchers began with a series of *in vitro* studies where they used containers full of fluid extracted from the cattle rumen. One study found that biochar lowered methane production by 12.7 percent. When supplemental nitrogen was added (the feeds used were low protein), methane production was reduced by up to 49 percent.[4]

The researchers followed up their *in vitro* work with a four-month *in vivo* feeding trial with young tropical cattle. Biochar was added as 0.62 percent of the dry matter (DM) fed. Two types of nitrogen were supplemented to the low-protein cassava diet. Biochar was responsible for a 25 percent increase in live weight gain without causing any change in DM intake, and a 22 percent decrease in methane emissions. Additive effects of biochar and nitrate combined reduced methane emissions by 41 percent.[5]

The reduction in methane emissions could be explained by the mechanisms described in the *in vitro* studies—the proliferation of methanogens. Or there could be other mechanisms that might account for both reduced methane and increased feed conversion. The researchers hypothesized that the increases in weight gain could result from biochar improving "microbial habitat where microbial consortia can come together for mutual benefits and efficient use of each of their metabolic end products." The right combination of bacteria can utilize more of the H_2, "producing more reduced end products of feed break down such as propionate." The result would be less formation of methane.

Finally, the researchers also looked at differences in the rumen fluid from animals both with and without biochar in their diet. They found that rumen fluid from biochar-fed animals produced less methane in an incubation than the rumen fluid from non-biochar-adapted animals. This indicates that the biochar-adapted animals had different consortia of microbes in the rumen microbial community that were responsible for the reductions in methane emissions.[6] This result was completely opposite to the effect of the pharmaceutical chemicals on the rumen, where the rumen microbes adapted to the presence of drugs over time, decreasing the effectiveness of the drug.

If the rumen environment adaptation response to biochar is to create a more efficient environmental community with less waste and lower methane emissions, then it contrasts with the antibiotic pharma approach that ends with the adaptation of the rumen microbes to eventually tolerate the poison. Biochar is the probiotic approach.

Since the Laos tropical cattle study was published in 2012, there have been dozens of studies *in vitro* and a few studies *in vivo*—with live animals. The live animal studies are very difficult and expensive to do because cows are large, valuable animals. They need to be housed in special, enclosed stalls in order to measure their methane emissions. As a result, the studies have been limited to only a few animals and only a few different combinations of feed ingredients and biochar. A review of both *in vitro* and *in vivo* studies on biochar in cattle feed found that overall the results are inconclusive but promising for both reducing methane emissions and improving feed efficiency.[7]

For the most part, the studies have not examined the use of probiotics that might be compatible with biochar and improve its functioning in the rumen, but recent work from researchers in Iran demonstrated that combining biochar with a *Lactobacillus* and yeast probiotic was effective at reducing methane emissions from young calves and also improving their health. The combination of biochar and probiotics was more effective than either one alone.[8]

Using more biochar might also improve the results. An *in vitro* study that tested different rice-husk biochar levels on rumen fermentation found that methane production decreased linearly with increasing biochar inclusion from 1 percent of the feed ration to 7 percent. The higher rates over 5 percent reduced feed digestibility, and the study's authors concluded that the best feed-inclusion rate for maximum benefit was 3 percent.[9] Most of the biochar feed studies have used feed-inclusion rates of 1 percent or less.

We Have Always Eaten Charcoal

Clearly there is a lot more work to do to understand how biochar could function as an animal feed supplement, especially in combination with probiotics and other ingredients. However, there is also a large compendium of research that was done in the past, before biochar became "trendy," that looked at charcoal or activated carbon as feed ingredients for animal health. In fact, this history goes way back before modern science.

Biochar or charcoal has a long history of medical use in both people and animals, primarily as a poison control and to cure a variety of digestive upsets.

The first written accounts of charcoal as medicine are found in Egyptian papyri dating to around 1500 BC. Hippocrates, Pliny, and Galen all wrote about the curative properties of charcoal for digestive ailments, wound healing, and other uses.[10]

Animals in the wild and in the barnyard have been observed self-medicating with charcoal. Colobus monkeys eat charcoal to help them digest leaves that have high levels of tannins. The leaves are otherwise nutritious and high in protein but the tannins interfere with protein digestibility, and charcoal binds the tannins. There is a funny video from BBC called "Colobus monkey medicine" that shows these monkeys stealing charcoal from a charcoal maker on his way to sell the product in town. The man was not amused.

A few years ago, I had advised some new arrivals to our countryside about how to make biochar for their garden. They left the biochar they made in a pile near the forest, but when they came back in the spring to get it, they found it was almost all gone. "The darn cows ate it," they told me. I had the same thing happen in my yard when deer consumed most of one of my biochar piles. Perhaps the best way to feed animals biochar is to feed them *ad libitum*—as they will. Animals often know what they need.

A textbook on animal husbandry dating from 1906 observed: "Swine appear to have a craving for what might be called 'unnatural substances.' This is especially true of hogs that are kept in confinement, which will eat greedily such substances as charcoal, ashes, mortar, soft coal, rotten wood, etc. It is probable that some of the substances are not good for hogs, but there is no doubt that charcoal and wood ashes have a beneficial effect, the former being greatly relished."[11]

Nineteenth-century and early-twentieth-century agricultural journals have many discussions of the benefits of various "cow tonics," mostly composed of charcoal and a variety of other ingredients that could be called spices, such as cayenne pepper, but also including digestive bitters like gentian. Manufacturers of these tonics claimed they would reduce digestive disorders, increase appetite, and improve milk production.[12]

Elmer Seth Savage, a professor of animal husbandry at Cornell University, recommended charcoal as a regular feed supplement to increase milk butterfat content and cow health, especially for dairy competitions. In his book *Feeding Dairy Cattle*, he reported that: "The grain mixture fed during the test consisted of 100 pounds of distillers, dried grains, 50 pounds of wheat bran, 100 pounds of ground oats, 100 pounds of hominy, 100 pounds of cottonseed meal . . . Charcoal is seldom if ever left out of the test ration by many of the breeders."[13]

Cow's milk was tested for butterfat content in competitions and top-producing cows could win prizes. Savage lauded the prizewinning Minerva Beets, a five-time champion dairy cow, still producing record-breaking milk at the age of fourteen

Figure 9.1. Carnation erected a monument to their champion milk cow. The plaque reads, in part: "Segis Pieterje Prospect. World's Champion Milk Cow. Born 1913. Died 1925. Her queenly worth deserved the gratitude in which this tribute is erected by her owner Carnation Milk Farms—1928." Photo by Ian Dewar / Alamy

years. The life of this healthy animal contrasts sadly with the fate of modern dairy cows, who rarely endure more than two years of milking before being relegated to hamburger. Another prizewinning dairy cow was honored with a monument at Carnation Milk Farms in Washington State. Segis Pieterje Prospect was a world champion cow who produced a record-breaking 37,000 pounds of milk in one year. Her dedicated milker, a man named Carl Gockerell, gave her the nickname "Possum Sweetheart." Savage reported that Gockerell kept meticulous records of her feed ration, noting that he fed 1 pound of charcoal for every 100 pounds of grain. Possum Sweetheart lived for twelve years under the care of Carnation Farms, where the rule was to treat all cattle, young and old, with patience and kindness. Signs that read "no swearing" were placed in the milking barns to help keep a peaceful environment for contented cows.

Biochar Rx

A variety of studies have looked at the impact of charcoal or activated carbon on animal health. Several studies further demonstrate the impact of biochar or charcoal on specific disease or toxic conditions, as summarized below. In most cases, the mode of action was sorption of a toxic compound or immobilization of a bacterium, or both.

ACIDOSIS

Acidosis is a chronic nutritional disorder in feedlots where cattle are transitioned from forage diets to grain diets. Acidosis is caused by explosive growth of *Streptococcus bovis* when grain is suddenly introduced. *S. bovis* produces large amounts of lactate, acidifying the rumen. It can kill the cow. Acidosis is not one disorder but rather a continuum of degrees. Effects of acidosis can manifest as a small reduction in feed intake or the death of an animal. According to the *Merck Veterinary Manual*, "administration of activated charcoal (2 g/kg) is believed to protect the ruminoreticular mucosa from further injury by inactivating toxins."[14]

One study also found that charcoal can help prevent acidosis. Sheep were fed two diets, roughage-based and concentrate-based. Activated charcoal (AC) was added at 0.3 percent of dry matter to the diet. The AC had no marked effects on feed intake, daily gain, or feed conversion, likely due to the low levels of AC provided: "However it was observed that the animals provided with AC in the concentrate diet did not suffer from diarrhea and easily adjusted to high concentrate feeding."[15]

BOTULISM

Botulism is a recurrent disease of cattle caused by the *Clostridium botulinum* bacteria. The bacteria can multiply in bad silage, and large outbreaks have occurred in the dairy and other livestock industries.[16] Cattle sicken from ingesting the toxins produced by the bacteria, and activated charcoal is a recommended treatment option for sorbing the toxins.[17]

Swiss and German researchers conducted an experiment in a dairy that was experiencing a number of common health problems, including reduced performance, movement disorders, fertility disorders, inflammation of the urinary bladder, viscous saliva, and diarrhea. Animals were fed different combinations of charcoal, sauerkraut juice, and humic acids over periods of four to six weeks.[18] Experimenters found that oral application of charcoal, sauerkraut juice, and humic acids influenced the antibody levels to *C. botulinum*, indicating reduced gastrointestinal neurotoxin burden. They found that regular feeding of charcoal and other supplements could have a tonic effect on cow health.

E. COLI

Cattle digestive systems are a reservoir of *E. coli*, including strains like O157:H7 that are pathogenic to humans. Feedlot and dairy systems that shift feed from forage to grain cause acid-tolerant strains of *E. coli*, like strain O157:H7, to predominate over less-acid-tolerant nonpathogenic strains. Fecal shedding of *E. coli* O157:H7 in cattle can infect water and soil and also result in carcass contamination, spreading the disease to the human food supply.[19]

Biochar can potentially reduce the spread of *E. coli*. An *in vitro* study added activated charcoal to a nutrient broth inoculated with *E. coli* O157:H7. The AC was effective in sorbing both the bacteria itself and the toxin it produced. The study also tested AC sorption of normal intestinal flora and found that the AC showed lower binding capacity to these organisms.[20]

Biochar also has potential to reduce the spread of *E. coli* in water and soil. Several studies have found significant reductions in transport of *E. coli* through soils that contained biochar.[21] Since *E. coli* infection within a herd seems to spread through water troughs, adding biochar to water troughs should also be studied.

PARASITES

Protozoal parasites are a major cause of diarrhea and loss of young calves. Researchers in Japan tested a novel compound of activated charcoal containing wood vinegar liquid as a treatment for infection with *Cryptosporidium parvum*. They found that the charcoal sorbed the oocysts and the wood vinegar killed the oocysts *in vitro*. Live calves were then infected with oocysts and half the calves were given the charcoal and wood vinegar treatment. Those calves recovered quickly while the control calves developed severe diarrhea.[22]

PESTICIDES

Pesticide residues of all kinds are present in animal feed. Biochar has been shown to sorb a number of pesticides and herbicides. Activated carbon has been used specifically for the purpose of removing organochloride pesticides from cattle. A study from 1971 reported at the conclusion of a trial that examined various alternatives: "The method that is effective as an antidote for pesticide poisoning in cattle is a combination of activated carbon and phenobarbital feeding. This method proved successful in a large-scale field trial involving 105 lactating Holstein cows that had been contaminated with aldrin."[23] Researchers in India found that feeding charcoal to dairy cows could reduce the excretion of organochloride pesticide residues in milk.[24]

TOXINS IN FEED

Mycotoxins are secondary metabolites of fungi and are a common contaminant of cattle feed, both grain and silage. Fungal growth can start in the field before harvest or develop in storage due to poor silage or inadequately dried grain. Charcoal and activated carbon can sorb these mycotoxins effectively and prevent toxic impacts on animals.

Researchers fed two groups of cows with contaminated silage or clean forage and supplemented both groups with 0, 20, or 40 grams of activated carbon per day. Cows on contaminated silage that received AC tended to improve intake and had higher butterfat content in milk compared to those that did not get AC.[25]

Ochratoxin is another toxic compound that is commonly produced by two species of fungi, *Penicillium verrucosum* and *Aspergillus ochraceus*. Adding 10 percent charcoal to the diet of pigs decreased the concentration of ochratoxin in blood, liver, kidney, spleen, and heart by 50 to 80 percent.[26]

Charcoal and Plant Secondary Metabolites

Animals that are fed charcoal on a regular basis are likely to experience better health and digestion as a result of charcoal's ability to neutralize pathogens and toxins. This will naturally result in better weight gain and feed conversion. Another potential benefit of feeding charcoal is the mitigation of Plant Secondary Metabolites (PSMs) such as tannins that are present in many forages.

Tannins are a complex and highly variable group of compounds that have some benefits and some detriments to ruminant digestion. Tannins tend to bind protein in the rumen, but that can have benefits to weight gain, as some tannins will then allow protein to pass through the rumen to be digested in the hindgut where it has more nutritional benefit to the animal because it has not been degraded for energy to feed bacteria in the rumen. However, not all tannins have this effect. Tannins are often present in high-protein forages such as legumes and their strong taste can put animals off their feed, lowering weight gain. Several studies have looked at the impact of supplementing with charcoal to counteract the effects of tannins and found that animals on a high-tannin diet that were fed charcoal ate more and gained more weight.[27]

Tannins also have the potential to reduce methane emissions. An *in vitro* study found that tannins from tropical tree leaves could suppress CH_4 production in rumen fluid without adverse effects on digestibility. The authors hypothesized that "methane reduction without significant reduction in VFA [volatile fatty acids, the product of rumen fermentation that actually feeds the animal] in most of the samples in our study was the result of the direct effect of tannin on methanogens."[28]

Essential oils are another group of PSMs that have been used in cattle feed supplements for benefits to health and feed conversion. Some researchers think they have good potential for reducing CH_4 emissions as well, and should be more thoroughly researched.[29] Biochar could be combined with essential oils or other PSMs and tested to see if they have additive effects on weight gain or CH_4 reductions.

Pets and Friends

Biochar used as a cattle feed supplement is both an old and a new idea. Traditional uses give us confidence that biochar is safe and effective to use as a feed supplement for animals of all kinds, including our pets and ourselves. A number of companies manufacture and sell charcoal feed supplements for animals and various health and beauty products for humans. To learn more about how to use charcoal for your own family's health, I recommend the book *Charcoal Remedies* by John Dinsley.[30]

John Dinsley is a world traveler who embarked on the hippie trail years ago and ended up working at a medical mission in Central America where he began his journey learning about the uses of charcoal in medicine. For many years after that, he traveled the world as a medical missionary working with groups establishing health clinics in remote areas. During those years he had many occasions to help people with his medical knowledge of simple natural remedies when no doctors or nurses were present.

Dinsley healed dysentery and food poisoning with charcoal taken internally. He healed spider bites and snake bites with charcoal poultices. He found that charcoal applied over deep wounds and abscesses could fight the infection and speed the healing. He shares a hundred or more stories of his experiences in the book.

Dinsley eventually came back to the United States and started his Charcoal House business to formulate and sell a variety of charcoal products for the health of humans and animals. Dinsley says:

> Every home should have charcoal on hand. It can be used as an antidote for poisoning and as a cleansing agent for our families, our animals and our homes. It can also serve to benefit the water, the air, and the land we take so much for granted. In short, it can help us sweeten and refresh this awesome world our Creator has so graciously given us.[31]

CHAPTER TEN

Building Soil

While Justus Liebig may have overthrown the humus theory (the theory that plants get their carbon from soil rather than from atmospheric CO_2) in the mid-nineteenth century, apparently it needs overthrowing again. The issue is the stability of humus. Does it really last for decades to centuries in soil? This has implications for schemes to store carbon in soil to keep it out of the atmosphere.

Related to the question of the stability of humus are questions about how humus is formed. Do microbes synthesize it? If not, then what causes humus to form in soil? What causes it to break down? What the heck is humus anyway?

What Is Humus?

Humus has been defined as the insoluble portion of soil carbon. If you scoop up some rich garden soil and shake it in a jar of water, after all the particulates settle out you will find some amount of brown tint to the water. That is the water-soluble fraction of soil carbon that is readily available for microbes to use in their metabolism. Some of the settled stuff is clearly the partially decomposed remains of plants. That is particulate organic carbon. Then there is a fraction that does not look like plant parts under a microscope, but can be dissolved in an alkaline solution and then precipitated out with an acid. That is called humic acid. What is left insoluble after that process is called humin. Similar kinds of carbon can be extracted from leonardite, which is mined from deposits of weathered, oxidized soft coal. This is the source of most of the commercial products called humic acid or humate—a salt of humic acid. This was known even back in Liebig's time when chemists did extractions with water, sodium hydroxide, and alcohol to characterize the different fractions of soil carbon. Despite the chemical manipulations, agronomists could not pin it down precisely. Soil scientists are still arguing about it today.

The gauntlet was thrown down in a 2015 paper by leading biochar researcher Johannes Lehmann and his co-author Markus Kleber in a paper titled "The Contentious Nature of Soil Organic Matter."[1] They say:

> Biological, physical and chemical transformation processes convert dead plant material into organic products that are able to form intimate associations with soil minerals, making it difficult to study the nature of soil organic matter. Early research based on an extraction method assumed that a "humification" process creates recalcitrant (resistant to decomposition) and large "humic substances" to make up the majority of soil "humus."

That theory is wrong, they say, because what we know as humus is simply a continuum of progressively decomposing organic compounds. They contend that the polymer-like, complex macromolecules of humus are actually artifacts of the alkaline extraction process that has condensed many smaller, easily degradable fragments of dead microbes and their metabolites into the humic acids. In reality, in the living soil, the humus portion is a diverse collection of organic molecules that are often easily degradable, and the only reason they exist at all is because they are protected from the ravages of hungry microbes by several distinct mechanisms:

Small, labile molecules are bound to metals and clays in soil aggregates.

Large, recalcitrant molecules like lignin are most easily degradable at the early stages of composting when there are plenty of small, organic molecules available to microbes as a source of energy to help them attack the chemical bonds. Later on, when the easy, fast food is used up, the microbes don't have the power to chew through lignin and other tough molecules.

Physical conditions including temperature, oxygen, moisture, and pH can speed up or slow down microbial decomposition.

Soil is a three-dimensional environment and just getting microbes in contact with the food they like is not trivial. Microbes have limited mobility in the soil ecosystem and must avoid being eaten by worms and other microfauna, hoping to find that rich cache of carbohydrates hidden behind a rock or tucked into a clay bed. They just cannot get to it all, and then they run out of energy and have to stay put on a biofilm and hope that a new flush of root exudate or dissolved carbon filters down to them. They are stuck in soil aggregates and colloids, and as more different species pile on, competition complicates things.

It's like the Serengeti Plain down there in a healthy soil, with teeming herds of living biomass. But instead of sitting in our Land Rover with a telephoto lens, all we can do is scoop up a whole valley, put it through a blender, and try to extract different solutions out of it. True, we can do more now than count the elements—we can analyze genes and enzymes to get an idea of the species and what they are eating—but we still can't keep watch around the water hole to understand the subtleties of the interactions.

In a later paper, Lehmann and coauthors expand on the role of biological diversity as a major factor in protecting soil carbon from degraders and consumers.[2] The more diverse the crops and ecosystems aboveground, the more diverse are the organic molecules in the soil. Every type of molecule needs a specific enzyme to break it down so microbes can consume it. Enzymes take energy to produce. In highly diverse systems, there will be many molecules that are not consumed for lack of energy. In a biologically diverse ecosystem like a natural forest, grassland, or wetland, carbon accumulates, creating the rich, black, virgin soils that we gave the name "humus."

Whatever it is, we know we need humus in soil. Even Justus Liebig conceded that humus was needed to help hold on to the soluble chemical nutrients he used to feed plants, lest they wash away in the rain. Humus is electronegative, like biochar, giving it high cation exchange capacity. Humus has staying power and it builds soil and keeps it healthy and moist. We really cannot do without it, but industrial agriculture has been destroying humus nonstop since Liebig's day. As discussed in chapter 1, Liebig did not foresee this. He did not understand that providing artificial nitrogen to soil would cause microbes to break into the humus stores in soil and consume them for their carbon, leaving us with weak, dried-out, compacted, disease-prone soils.

Some scientists would like us to abandon the term "humus" in favor of the "soil continuum model," but I think that would be a mistake. First of all, not everyone agrees with all of their conclusions.[3] And while it's great for researchers to create specialized terms that help them make necessary distinctions, for those who are working the soil, we can continue to use the term humus to describe the entire range of soil organic carbon.

I also want to retain "humus" for cultural reasons. The word comes from the Greek *choma*, meaning ground. We are humans, *Homo sapiens*, from the same root word. As humans, we are nothing without humus. We will be much better off if we can humbly acknowledge our utter dependence on healthy soil, and that humus is critically linked to soil humidity. On average, every increase of 1 percent in soil organic matter holds an additional 26,000 gallons of water per acre. When it comes right down to it, humus is more valuable than gold.

We moderns lack the humility to recognize that we all come from the ground, and the power of Nature is greater than we can ever fully understand. Our arrogance is quite literally killing us, and that is humiliating. We need an attitude adjustment as we face the future and all its problems. We will be OK if we can stay engaged with our humanity, working together with others to restore soil humus, and we will find that a good dose of humor can help a lot.

When does a farmer need to add pita bread to soil? *When it is hummus!*

How Can We Help?

One of the most important points that Lehmann's group makes is that the goal should not be to just stash carbon away in soils where it won't be used by life: "Functional complexity as a driver of soil carbon persistence suggests soil management should be based on constant care rather than one-time action to lock away carbon in soils."

The emphasis should be on flows, not stocks. Keep the carbon moving to keep it in the living biomass. It is life itself that is the best "carbon pump" to remove carbon from the atmosphere and house it in the living bodies of organisms at all scales. The more complex and diverse life is, the more carbon it sucks from the sky.

Lehmann's group provides a list of recommendations for regenerative soil carbon practices, summarized as:

- Where possible, convert land to perennial vegetation.
- Restore wetlands and avoid draining or disturbing them.
- Grow a variety of different nitrogen-fixing plants and cover crops.
- Reduce tillage and soil disturbance.
- Rotate crops and overlap their growth cycles so roots are always in the ground.
- Mix perennials and annuals, as in agroforestry.
- Add soil-building amendments including a variety of residues, manures, compost, and biochar.

We cannot just add biochar and walk away. But biochar is very helpful in a broader program of soil regeneration that supports the soil microbiome and diversity. As we have seen, biochar is an extremely effective substrate and habitat for microbes. Biochar carbon also balances the excessive nitrogen applications that are part of our agriculture.

Balancing carbon and nitrogen in compost and soil is a key factor in building humus and healthy soil. When the C:N ratio is optimized for soil denizens, plants get fed, and residues become humus. Humus is the rainy-day larder or, as biodynamic farmer Hugh Lovel termed it, "the humus flywheel."[4] It's there to take up the slack when nutrient inputs slow down. When rain washes out soil solution nutrients, hungry microbes can still find something to eat. But when soil solution is constantly recharged with chemical NPK, C:N ratio decreases.

Microbes attack the humus just to get at the carbon. Carbon is burned out of the soil and the nutrient larder along with it. The result, as we see, is that agricultural soils worldwide are severely depleted of carbon.[5]

Carbon also needs to be balanced and used appropriately. Adding too much labile carbon like straw or sawdust immobilizes nitrogen temporarily, as microbes chow down and store the available nitrogen in their tissues, robbing the crop that you planted. I made the mistake of adding a sawdust mulch around my strawberries one year, thinking they would like the acidity. My strawberries were terrible! Some kinds of biochar can also immobilize nitrogen, such as low-temperature biochar that has a significant content of labile carbon. Biochar made at the high temperature of a flame-cap kiln will not have this problem to any great extent.

The more likely problem of adding raw, uncharged biochar to soil is that it will adsorb nitrogen. This is actually a good thing as it tends to support the nitrifying bacteria that convert ammonia to nitrate. Nitrate is an anion that is generally not held by clays and organic material that have cation exchange capacity. Nitrate is easily leached from soils with rain, but biochar seems to be able to hold on to it pretty well and keep it available in soil as plant nutrition.

Even biochar that has an alkaline pH will still do a good job of retaining nitrogen. Despite some ammonia escaping due to alkalinity, much of it is still captured by biochar.[6] We saw this result in our experiment at Michaels Ranch, adding a high-ash, high-pH biochar in the winter feed barn. The biochar pile still managed to capture more nitrogen than the untreated manure pile.

There is one case where you can usually just add biochar and walk away: when planting legumes. Legumes fix their own nitrogen and biochar is a big help in that process as it supports the nitrogen-fixing bacteria that form nodules around the roots of a legume. A meta-analysis that looked at ninety-six different studies of biochar in legume crops found that biochar significantly increased the number and weight of nodules, and plants were more productive.[7]

In every other case, in order to avoid potential problems with adding fresh, reactive biochar to soil, we have an easy solution: compost it. Compost is a way to prebalance the C:N ratio of amendments before you add them to soil. If you want to preserve humus, make sure everything you add to the soil has a balanced C:N ratio. This is why biochar is the perfect pairing for manure.

How to Use Biochar in Compost

The ideal starting C:N ratio of a thermophilic compost pile will be about 30:1. This is the balance of browns and greens that make up an active compost. The C:N of the finished humus will be about 15:1 to 10:1. For reference, the C:N of a living microbe is around 8:1, so high-quality finished compost still has plenty of usable energy for microbe metabolism.

The composting process is governed by various physical parameters: carbon and nitrogen content of the ingredients, aeration, moisture content, temperature, bulk density, pH, and the absorption capacity of bulking agents. Proper attention to all of these factors will usually result in high-quality compost, but biochar changes some of the rules.

Compost moisture, aeration, and turning are all affected by biochar. Biochar holds water in pore spaces and voids, and in the spaces between particles. Biochar also holds air in these same spaces. Using biochar as a bulking agent to aerate compost has several advantages over bulking materials such as straw or wood shavings. One function of a bulking agent is as a structural support to create air voids between compost particles. As the compost ages, the bulking agent degrades and will collapse and densify, reducing aeration. Biochar prevents this.

I find that in my garden compost pile, I no longer need to turn it at all, but it is a small pile, yielding about 2 cubic yards (1.5 m^3) of finished compost each year. A few years ago, I visited Missouri Organic Recycling in Kansas City. They compost all the organic waste from Kansas City in a massive facility with many composting windrows. After they started using biochar, they were able to reduce the amount of turning. The compost also finishes in half the time. That means that the company can process more compost without having to enlarge their facility. I asked them how they got started using biochar. CEO Kevin Anderson said he had been approached by some organic growers about the quality of the compost. They were not happy with it. They introduced him to David Yarrow, a gardening expert, who showed Kevin how to use biochar,

Figure 10.1. Kevin Anderson talks about the success Missouri Organic Recycling has had with using biochar in their compost operations.

rock dust, clay, and a few other ingredients to improve the compost. Kevin found that biochar really helped with odor reduction. He said, "The smell of ammonia is the smell of nitrogen loss. Our Green Frontier compost has more nutrients and greater stability according to the tests we run that are approved by the US Composting Council. And the organic growers really like it. Some of them won't use anything else."

David Yarrow has been an amazing teacher to many in the biochar world. You can find his useful article published online in Eco Farming Daily: "Biochar, Prepping It for Soil."[8]

We have also learned that adding biochar to compost must be done with careful reference to C:N ratios. For best results using biochar in compost, it is beneficial to combine biochar directly with a high-nitrogen source and then add that to compost with other materials that have balanced C:N ratios. Here is a list of different considerations for using biochar compost:

- C:N of biochar itself could be about 100:1 or greater—it depends on the biochar.
- Typically, only about 5 to 30 percent of the total carbon in biochar is available for microbes to consume. High-temperature biochar made in flame carbonizing process will be on the low end of that scale.
- Biochar can adjust the C:N ratio of high-nitrogen ingredients like manure, mainly by holding on to the nitrogen so it is not immediately available to microbes. Plan accordingly so your other compost ingredients are balanced to match it.
- Biochar does not have a lot of degradable carbon. If you don't have enough degradable carbon in greens or manure, you may need to add more degradable carbon, such as sugars, cellulose, or carbohydrates, along with the biochar, to help feed microbes and generate thermophilic composting conditions.
- Biochar increases nitrogen retention in finished compost as ammonia converts to nitrate.
- Biochar tends to promote a more fungal-dominated compost.
- Biochar improves compost maturity and humus content.
- Biochar content for good compost ranges from 3 to 25 percent depending on nitrogen content of manure or green waste, the amount of degradable carbon, and other ingredients.

What happens to degradable carbon in compost with biochar is very interesting. I once added biochar to some washed dairy manure from a lagoon that was mostly fiber low in nitrogen. When I added biochar to one pile, it cooled down by about 5°F compared to the unamended pile. I then drenched it with some molasses (degradable carbon) dissolved in water, and the pile started heating up again—just like throwing more fuel on a fire. I think what happened is that the sugar gave the microbes the energy they needed to go after the nitrogen retained in the fiber and the biochar.

There is a lot to learn about using biochar in compost, so let's explore some different techniques for improving compost with biochar, along with methods for evaluating the results.

Preserve Nutrients with Effective Microorganisms

When you have very high-nitrogen ingredients like food waste, urine, grass clippings, and fresh manure, add biochar as the waste is generated to preserve nitrogen that wants to volatilize. In my vegetable production system, I use bokashi to pickle and preserve nutrients from my food waste in a precomposting process before I add it to the compost pile. I have already shared my method for making a biochar urinal to capture and preserve nutrients in urine. Here is my recipe and procedure for handling food waste:

PICKLED KITCHEN WASTE

Kitchen waste can go in your outdoor compost pile, but it may attract pests and cause other problems, especially if you have dairy or meat scraps. Processing it first in a bokashi bucket is the way to go. You can keep the bucket under the sink in the kitchen, in the garage, or outside on a porch or balcony. Pickling kitchen scraps reduces odors and preserves the nutrients in the food scraps.

Ingredients

1 gallon (4 L) dry biochar
½ gallon (2 L) dry bokashi

Procedure

Mix together half of the biochar with the bokashi. Put a couple of inches (5 cm) of the remaining biochar in the bottom of a 5-gallon bucket. Make sure it is as dry as possible because it needs to soak up all the liquid from the food waste. Some advocates of using bokashi with food waste recommend using a bucket with a drain to remove the liquid periodically, but biochar can just absorb it. If the food is really wet, add more biochar.

Start adding food scraps as you generate them. Whenever you add a couple of inches of food scraps, sprinkle about a cup of bokashi/biochar over the top. Use a board to tamp it down and remove air from the waste. Soon you will see

Figure 10.2. A. Screening the previous year's finished compost. B. Building the new pile with a barrel of pickled food waste. C. Adding more layers of green weeds, biochar, and rock dust.

a white fuzz growing on top. This is a beneficial, filamentous bacteria called actinomycetes. Keep the lid on tight, and when the bucket is full, you can either dump it into an ongoing compost pile, or add it to a larger barrel like a 58-gallon plastic pickle barrel and keep it until it is time to construct a new compost pile. The pickled nutrients will be preserved. This is a much more potent fertilizer than you get if you just toss food scraps in an open compost pile where they mostly dehydrate and lose nutrients.

My process is to keep it in the pickle barrels for a year until I start a new compost pile in the spring. I layer the goopy stuff with browns and greens from spring garden clean up, along with rock dust and some additional biochar. I find this works very well to make a rich, trouble-free compost. Once I build the pile, I don't turn it all. I keep it moist and cover it in the winter when it rains. In the spring, I harvest and screen it and add it to my garden beds when I plant them.

When a garden or a farm has large amounts of sugary waste, fermentation with *Lactobacillus* and biochar is the best option. Agriculture consultant and bokashi expert Cuauhtemoc Villa joined me for a biochar workshop at an Oregon vineyard to demonstrate this. The vineyard had a long windrow of decomposing grape pomace. Cuauhtemoc spread a layer of biochar on top of the pile. He sprayed 100 gallons or so of EM-1 solution on the char and then invited workshop participants to mix it well with shovels. In a few weeks it would turn into a rich soil amendment with stabilized nutrients.

While I have been inspired by different systems described by Teruo Higa, inventor of Effective Microorganisms, I recently came across the SPICE method developed and promoted by Australian Gerry Gillespie. SPICE stands for "static pile inoculated compost extension." The SPICE method is appropriate for large- or small-scale composting. Gillespie provides a wealth of information about it on his website at GerryGillespie.net. Gillespie explains that the method is based on the "Indore" process of Sir Albert Howard, a leading British agricultural scientist in the early-twentieth century. It is a two-stage process, starting with an aerobic phase and then switching over to a low-oxygen, fermentation stage.

If you do not have access to EM-1 culture, any good *Lactobacillus*-based culture will work, including a homemade serum using indigenous microorganisms or a commercial silage inoculant. *Lactobacillus* and biochar are a winning combination. Researchers in China composted sheep manure and wheat straw with biochar and EM-1.[9] They found an increase in fungal degraders in the compost resulting in higher humus content in the finished compost and lower greenhouse gas emissions during the process. Fungi are important for humus production because they can degrade cellulose and lignin and get them started on the soil carbon continuum process that ends with the formation of humus.

Figure 10.3. Transforming a putrid pile of decomposing grape waste into rich compost. **A.** First spread a layer of biochar. **B.** Spray with EM-1 inoculant. **C.** Mix well.

Experimenting with Biochar Compost

Because biochar has some unexpected impacts on the C:N ratio of compost, and because everyone has different kinds of materials available for composting, it may be helpful to do some small-scale experiments first to determine optimum combinations for producing good compost using biochar.

First decide what you want to test. You need a hypothesis: Biochar will impact compost processes (time and temperature) or compost qualities (maturity, nutrient content, pH, humus content, and fungal to bacteria ratio). Read through these ideas for methods and measurements to help you design and monitor your experiment.

Procedure

1. Set up one compost with biochar and one compost without biochar (a control treatment). Or set up multiple treatments with different rates of biochar addition. If you like, you can replicate the experiment with multiple piles of each treatment.
2. Pulverize biochar to a size that is mostly ½ inch (1 cm) or smaller. Some larger pieces are OK and will help with aeration.
3. Try to determine the C:N ratio of all of the ingredients, keeping in mind the special qualities of biochar. You can find C:N calculators online to help you with the math.

Figure 10.4. Make an experimental compost to test your combination of ingredients. Use bins, open piles, or 100-gallon compost sacks. Measure your ingredients and mix by layering.

4. Apply biochar at between 3 percent and 25 percent by volume to the biochar compost. Use more if you have a high-nitrogen content, and less if you have a lot of browns.
5. Choose a compost container: wood or wire bins, open piles, or fiber sacks.
6. Obtain a compost thermometer with a long enough probe to reach the center of your piles.
7. Construct piles in layers, taking care to add bulk materials and biochar in enough layers to allow natural aeration.
8. Water both piles the same.
9. Monitor moisture—the compost should pass the "squeeze test." Take a handful of material and squeeze. No water should come out, or maybe just a drop or two, but when you open your hand, there should be enough moisture to hold the material together so it does not fall apart instantly.

Measurements

- Make a data sheet to record the date, time of day, ambient temperature, and temperature measurements for each pile.
- Take daily temperature measurements for the first two weeks.
- Take weekly temperature measurements for next four weeks.
- Sample compost after ninety days and send for testing.

Lab Tests

Temperature measurements will tell you how the composting process compares with and without biochar, but to know how it compares as a fertilizer, you will want to test the results. You should expect to see greater retained nitrogen in the biochar compost. You may also see a difference in the indicators for compost maturity. Below are some tests you can request from most soil labs:

- Nutrient profile
- C:N ratio
- Compost stability test (CO_2 respiration)
- Phospholipid fatty acid (PLFA) analysis to indicate the composition of the microbial community (some labs can do this)

I sent a sample of my regular garden compost to Ward Laboratories in Kearney, Nebraska, for a PLFA analysis. These fatty acids are part of the cellular membrane of microbes and the analysis provides a snapshot of the amount and proportions of fungi, bacteria, and protozoa. I was especially curious about the ratio of fungi to bacteria. A higher ratio is an indicator of better-quality compost. Table 10.1 has the results.

Table 10.1. Analysis of Wilson Biochar Garden Compost by Ward Laboratories, Inc.

Indicator	Value	Rating
Total microbial biomass	2601 nanograms/gram	slightly above average
Functional group diversity index	1.6	excellent
Fungi:bacteria ratio	0.51	excellent
Predator:prey ratio	0.048	excellent

Except for the total biomass of microbes in my sample of mature compost (sampled in early spring), all of the other indicators were beyond the top of the scale for excellence. The predator:prey ratio is especially interesting. This is the ratio of protozoa to bacteria. According to Ward Labs, protozoa feed on bacteria, which helps release nutrients, especially nitrogen. A higher ratio indicates an active community that has sufficient nutrients to support the higher trophic levels that predators operate on.

Bioassays

You can use the seed germination test described in chapter 5 to test for compost maturity. Immature compost is still full of microbes actively secreting enzymes to attack and break down biomass. Germinating seeds are vulnerable in that environment. To learn even more about your compost, use a plant growth bioassay to compare soil with and without biochar compost to see what differences the composts make in plant growth. For instructions, see the appendix for the practice guideline: "Plant Bioassays to Evaluate Biochar Compost."

Figure 10.5. Don Morrison's biochar compost pile just after construction (*left*) and fully mature with fungi growing in it (*right*). Photos courtesy of Don Morrison

Table 10.2. Pot Trial Bioassay Results Comparing Two Kinds of Compost to Control

Treatment	Germination (%)	Secondary Leaves (count)	Biomass (grams)
Composted biochar	100	67	2.04
Fresh worm castings	84	57	1.87
Control	98	41	1.3

One of the most amazing experiments we ever did was to test biochar from my friend Don Morrison's compost pile that included chunks of biochar from a Ring of Fire kiln, goat manure, straw, and garden weeds. These were all layered in a wire bin and composted for several months.

We were interested in seeing how the composting process might have affected the biochar, so Don took a sample of the finished compost and sent it to me. I picked out some of the biochar chunks that were embedded in the compost.

I crushed the biochar—just the biochar with none of the compost—into fine particles and mixed about 2 tablespoons into a plain peat potting soil in each of several 4-inch (10 cm) plastic pots. I prepared another treatment with some fresh, wet worm castings and a third treatment for a control that was just the plain peat potting soil. I followed all of the instructions in the appendix and grew cucumber seeds for two weeks in a controlled environment. At the end of the growth period, I counted the number of seeds that germinated and the number of secondary leaves. I cut the plants off at the soil level and dried them and weighed them. The results astounded me.

Figure 10.6. A simple pot trial bioassay compares different soil amendments to see how they affect plant growth. From *left* to *right*: composted biochar, fresh worm castings, control.

The composted biochar had adsorbed nutrients from the compost and had a wonderful growth-stimulating effect. While the worm castings probably had plenty of nutrients, they also might have had a lot of fresh enzymes that would inhibit seed germination. The biochar not only had plenty of nutrients for plant growth but also seemed to absorb any harmful compounds that would inhibit seed germination.

Several researchers have reported that composting deposits a rich array of nutrients and minerals on the surfaces of biochar particles.[10] My pot trial confirms that these are available for plants to access.

I have used this bioassay to test peat-free soil mixes that I and others are trying to develop. We found that a proprietary mix of composted biochar and other ingredients far outperformed every commercial potting mix we put it up against. The bioassay is probably the most valuable test you can do because it asks the plants what they like and lets them tell you directly.

Chromatograms— Another Way to Look at Humus

A soil lab test can give you a number for the humus content of your biochar compost, but given what we know about the soil organic continuum, what does that really tell you? "Pfeiffer's circular chromatography" (PCC) was developed by Ehrenfried Pfeiffer, a student of Rudolf Steiner and a proponent of Steiner's biodynamic farming methods.[11] The technique uses an alkaline extraction of humus from a sample of soil or compost and a filter paper soaked in silver nitrate solution—the same chemical used in photography. The solution is slowly wicked onto the filter paper where it reacts with the silver nitrate and makes a picture.

Remember what Johannes Lehmann's group said about the difficulties of analyzing humus because the organic matter will "form intimate associations with soil minerals, making it difficult to study the nature of soil organic matter." Chromatography allows the soil to paint a picture of itself as the different components are pulled out onto the filter paper by capillary action, reacting with the silver nitrate along the way. The pictures are analyzed according to radial features, concentric features, and color. Pfeiffer said that deep-brown colors represented humus, and violet and gray colors indicated more mineral-bound components, especially in the central zone where heavier minerals would fall out of solution first. He took samples from many fields that had similar levels of chemical nutrients as analyzed by a soil lab but performed differently when growing crops. Invariably, the fertility differences could be distinguished in the chromatograms, where the fertile soil displayed a more complex pattern. He also compared chromatograms of composts at different stages of maturity and mineralization and could identify differences in the respective pictures.

Several studies have attempted systematic correlation between the features of the chromatogram and soil nutrient status and soil health, with mixed results. One paper concluded: "It can be suggested that radial pattern differentiation in PCC is a sign of high soil quality, whereas domination of concentric features indicates an unfertile soil."[12]

William F. Brinton of Woods End Laboratories, a leading developer of compost techniques and analysis, has a paper on the method.[13] Brinton thinks it is useful as a way to characterize the development of humus. The lab produced chromas for twenty different soils and found that they could distinguish aspects such as "mineral content, pH, mobility of humus, and the humic-fulvic acid ratio of the humus." The lighter fraction of fulvic acid migrates to the outer edges of the ring. In poor field soils with no recent organic matter inputs, this outer ring may be totally absent. In mature compost where humus is highly complexed with minerals, the color may be less saturated because the more recalcitrant humus is less extractable by the alkaline solution.

There is obviously an art to making and interpreting these soil pictures, but it's a fun science project and with experience, it could become useful, especially for visualizing the impact of biochar in soil and compost. I found the technique simple to do in my home workshop with inexpensive supplies: filter paper, silver nitrate, and potassium hydroxide, along with some 50-milliliter centrifuge tubes and plastic petri dishes. Thanks to Ken Avery for publishing a small book on the technique with very detailed instructions and a list of supplies, I was able to easily produce about twenty chromas from samples of composts and soil.[14] A report from Australia titled "Using Paper Chromatography for Assessing Soil Health in South Western Australia" also has a good set of instructions and an interpretation guide.[15]

I share a few of my chromas here to illustrate some of the examples I have already discussed. Figures 10.7A and 10.7B are from the cattle manure with and without biochar at Michaels Ranch, discussed in chapter 8. Figure 10.7A is manure scraped from the winter feed barn and piled with no other additives. Figure 10.7B is from the year where we added biochar to the winter barn in the fall. The cattle pooped on it all winter and then it was scraped and piled to compost. The second picture with its more developed radial features is a more fertile and biologically active compost.

Figures 10.7C and 10.7D compare commercial compost with and without biochar, from Missouri Organic Recycling. The first picture is their standard Naturewise compost without any biochar. The second picture is the Green Frontier compost with biochar added. The radial features are more developed

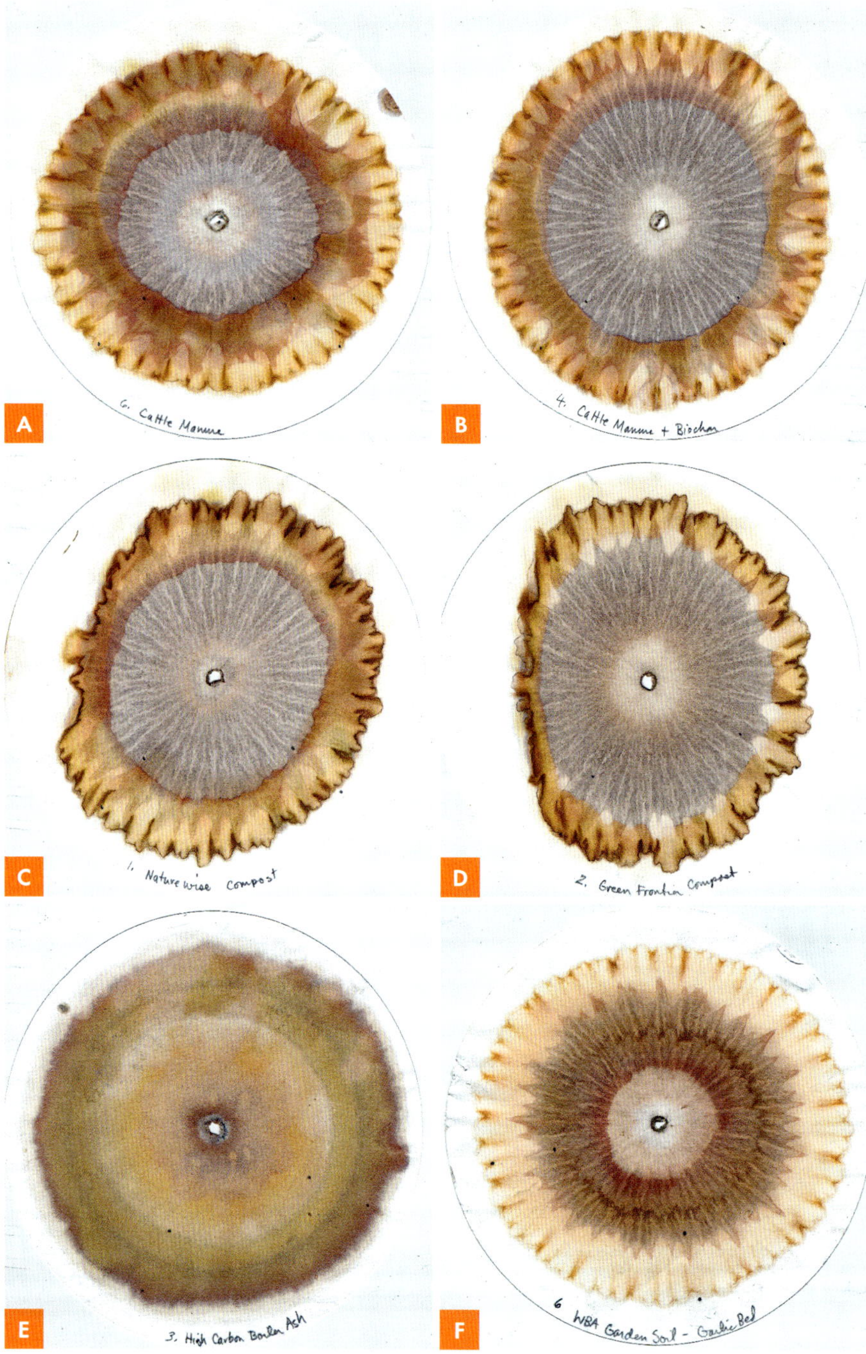

Figure 10.7. Chromatograms made from manure, compost, soil, and high-carbon ash. **A.** Plain cattle manure, scraped and piled. **B.** Cattle manure mixed with biochar in the barn. **C.** Naturewise (standard) compost from Missouri Organic Recycling. **D.** Green Frontier (with biochar) compost from Missouri Organic Recycling. **E.** High-carbon wood ash. **F.** Soil from a garlic bed in my garden.

Figure 10.8. My garlic crop.

in the Green Frontier. This would correspond to test results showing more nutrients and greater stability in the Green Frontier.

Figures 10.7E and 10.7F are from the high-carbon wood ash that we used as biochar in the Michaels Ranch cattle barn, and a sample of soil from a bed in my garden that was growing garlic at the time. You can see how amorphous and undifferentiated the carbon ash is. This is a purely chemical substance with no life. In contrast, the garden soil looks like a very fertile, biologically active soil. The radial features are strong, extending to the edges, indicating the complex organo-mineral compounds in the soil, and the color is a rich brown from the humus. I can attest that the garlic is great!

The Proof Is in the Pudding

Does biochar help build soil? In my garden it sure does! I have been gardening in the present spot for thirteen years now, gradually expanding from about 20 beds to 40 with a total raised bed space of 1,720 square feet (160 m^2). The garden is on a cobble terrace above the river with very little native soil. Over the years, I've had added about 30 cubic yards (23 m^3) of cow manure and some bales of straw mulch, but the rest is my homemade compost and biochar. The original beds were made by the previous owner and had no biochar. Every year, for the first few years I gardened here, I had to scrounge up soil and compost because the beds shrank so much. That is when I brought in all the cow manure. As I slowly

Figure 10.9. A tomato bed in my garden overflowing with biochar-enhanced soil.

added biochar to compost and fertilizer blends, the soil became looser, blacker, and more fertile. I am growing soil now, and every year I have more, allowing me to expand the garden. Now I am up against the edges of the garden fence. Oh, whatever shall I do with all this soil? If you would like to have my problem of overflowing garden beds, see the next chapter for some additional recipes and techniques for using biochar in your garden.

CHAPTER ELEVEN

Biochar Recipes

Here is a series of recipes and ideas for using biochar that I have either used myself or heard about and want to try. These recipes and guidelines are for ferments, composts, fertilizers, pest and disease control, potting mix, trees, and mulch. I also include some tips on how to apply biochar and resources for learning more.

Biochar in Ferments

To my mind, bokashi is the best ingredient to pair with biochar to make an amazing fertilizer on its own, and as an additive to improve compost and manure management. I have described how I use it in terra preta sanitation and manure management in chapter 8, and food waste composting in chapter 10. Here is the recipe I use for propagating my EM-1 culture, followed by two approaches for making bokashi.

PROPAGATING EM-1 CULTURE

I use a standard beer fermenter to brew my EM-1 culture. It may be helpful to use a clear container that will admit sunlight in the brewing process. I have been using the same culture as a starter for many years now, and it seems unchanged to me. However, experts warn that it may have devolved to being just a lactic-acid bacteria (LAB) serum with yeast, and that only the commercial EM-1 products have the full consortium of bacteria. Even if that is the case, the LAB itself seems to do the job I need it to do, which is to pickle my waste products: manure, urine, and food scraps.

To expand your purchased EM-1 culture, start with 1 liter of EM-1 and add 1 liter of molasses to 20 liters of water that is just a bit warmer than room temperature. You can add a small amount of minerals as rock dust or sea salt and a bit of biochar, perhaps one cup, but not too much because you don't want the alkalinity to interfere with the pH drop. Monitoring the pH of the solution is how you know when it is ready. Stir all the ingredients well and place a loose-fitting cover on the container. Keep it warm (room temperature is fine) and check back in two or three weeks to see if it is done. If the pH has dropped to 3.5, it is ready to use. I like to use high-quality pH paper rather than mess around with a lab-grade pH meter. It seems accurate enough for this purpose.

Figure 11.1. EM-1 culture brewing in a beer fermenter.

When it is finished, decant the liquid into smaller containers with very little headspace. You do not want the culture exposed to oxygen any more than necessary. Glass containers with tight-fitting lids are not recommended because the liquid may continue to release a small amount of gas. Plastic soda bottles are perfect because they can expand a little bit. Label the bottles and store them at room temperature. I have kept cultures for up to a year this way and they don't seem to go off. If you wonder about a culture, test the pH. If it is over 4, don't use it in soil, but toss it in the compost.

BEER MASH BOKASHI

I have a friend who brews beer. Periodically, he dumps a 10-gallon bag of spent grains at my door step. Here's what I do with them:

Ingredients

1 cup (240 ml) molasses (unsulphured)
1 quart (1 L) hot water
About 10 gallons (38 L) beer mash
1 cup (240 ml) EM-1 microbial inoculant
1 to 2 gallons (4–8 L) dry, crushed biochar, enough to absorb moisture and residual alcohol in the wet mash

Procedure

1. Mix the molasses in the hot water until it dissolves. Add cold water so it is about body temperature or a little warmer and mix with EM-1 inoculant.
2. Put the beer mash in a wheelbarrow or bin for mixing and sprinkle the liquid EM-1–molasses solution over it. Mix well. If you have a cement mixer dedicated to soil mixing, use that.
3. Start adding biochar to the mix, a quart or so at a time. The goal is to add enough biochar to soak up the liquid. You want the final mixture to pass the "squeeze test."
4. Once you have the material at the right moisture level, start packing it into 5-gallon buckets with good tight lids. I like to use the spin-on Gamma Lids because they are easy to get on and off, but snap-on lids are fine, too.
5. Pack the bokashi very firmly. I use my feet to stomp on it in the bucket. You don't want any air in it because this is an anaerobic ferment, like pickles or sauerkraut.
6. Ferment the material for two weeks or so. In colder weather, if the buckets are in an unheated space, it will take longer.

ECONOMICAL BOKASHI

If I am not using the free beer mash, I will buy 50 pounds of whatever is cheapest at the feed store. It could be wheat bran, rice bran, beet shred, or flaked barley. It can be any degradable carbon with some nitrogen. Joel Combot, a farmer I know in Manitoba, is using shredded cattails from his wetland. The best thing to use is whatever you have that is free or that you need to dispose of, like the beer mash. When I use a dry ingredient like wheat bran, I typically need to add a lot of water to it, but not too much. First add the molasses and EM-1, then add water slowly, 1 gallon at a time, as you mix the bokashi, until it passes the squeeze test. I like to put an inch of biochar in the bottom of the bucket to absorb moisture, but I don't mix much biochar in the fermenting material, because I don't want to inhibit the pH drop as the feed ferments. Instead, I mix the bokashi with biochar after it is finished fermenting.

Wet bokashi can stay in the buckets for several months, but it's best to use it sooner or dry it for long-term storage. When you open a bucket, you may see some green mold on the top where it was exposed to oxygen. Scrape that off and toss it. White fuzz is good—that is the actinomycetes bacteria, a helpful type of soil bacteria.

Wet bokashi can be used as a fertilizer. Mix it with biochar to mitigate the acidity. Bury it deep in the soil out of the root zone before you plant. It is quite acidic (pH 3.5 to 4.0) and that can be hard on young plants and seedlings. Best

Figure 11.2. Barney the Barn Cat is crazy for bokashi drying in the greenhouse.

practice is to add it at least a week before planting, but if you bury it deep, you can plant and let the roots grow into it when they are ready. The bokashi will have broken down into more pH-neutral forms by the time the roots reach it.

You can dry bokashi by spreading it out on a tarp to dry. Dry bokashi is very handy to have around for several uses. You can use it to "pickle" your kitchen scraps and other smelly things. It is also a probiotic feed ingredient for animals. My cat is crazy for it! When I had it drying on screens in my greenhouse, I could not keep him away from it.

Fermentation Experiment

I wanted to see if I could lower the pH of biochar using EM-1 inoculant only. I prepared a 5-gallon bucket of coarse biochar and added a cup of EM-1 inoculant with one cup of molasses and just enough water to wet the char. I sealed the bucket and stored it in a warm place for four weeks. After the fermentation, I added treated and untreated biochar to beakers with pH-neutral water and tested the solutions with litmus paper. The inoculated biochar was pH 6 as compared to pH 9 of the untreated biochar. I left beakers of treated and untreated biochar solutions in the open air for another two weeks and retested the pH. The fermented biochar was now pH 7, while the untreated biochar remained at 9. This demonstrates that we can successfully lower the pH of biochar using EM-1 inoculant.

Cultured Biochar

Sometimes you may end up with a lot of biochar that you want to use immediately, without waiting for it to compost. Or perhaps you don't have enough manure or grass clippings or other ingredients to make compost. In that case, you can make a cultured biochar that will condition biochar for immediate use in soil. I learned about this technique from biochar pioneer Josiah Hunt. Josiah has a lot of very useful biochar information you should be sure to check out on his website, PacificBiochar.com. To make cultured biochar, use EM-1 or another inoculant like indigenous microorganisms, and mix it into a pile of biochar with a food source for the microbes. This can be any kind of carbohydrate, liquid or granular, or something with more protein like fish hydrolysate. You don't need a lot of food to get it going. This is kind of like the reverse of regular compost. Instead of adding 10 percent biochar to compost, you add 10 percent degradable organics to biochar to make cultured biochar. You can mix minerals like rock dust with the char. When the weather is warm, make a heap of the mixed dry ingredients and make activated EM-1 by mixing twenty parts water to one part EM-1 and one part molasses. Pour on the inoculant, getting the pile thoroughly soaked. Then cover with a tarp and wait for two weeks. This treatment will lower the pH of the biochar, oxidize the surfaces of the biochar pores, and start moving in the microbes.

Sea Salt Biochar

Teruo Higa, the inventor of the EM-1 inoculant, has a recipe for culturing biochar with sea salt that would be interesting to try. Here is the recipe, adapted from his website, emrojapan.com:

1. Make charcoal (using a "smokeless carbonizer" like the Japanese cone kiln, which is a flame-cap kiln).
2. Quench with activated EM-1 made using seawater.
3. Add charcoal to salt in a ratio of 1:1 to 2:1.
4. Mix well.

Note: It is not necessary to let the sea salt biochar ripen. Put it in a fertilizer bag or container, and because salt tends to absorb moisture in places with high humidity, keep it in a well-ventilated place, such as a farm equipment storage shed where it won't be exposed to rain.

Dr. Higa uses this mixture as a fertilizer in the following way: "The amount of charcoal bokashi given to this flower bed is about 200 g (about a handful) per 1 m^2. Regarding application, I think it can be applied about once every two weeks to one month, while checking the condition of the plants."

Biochar Fermented-Liquid Fertilizer

I am very interested in Korean natural farming techniques, especially as detailed in Youngsang Cho's book *JADAM Organic Farming*.[1] I have been making liquid fertilizers somewhat haphazardly by adding green garden weeds to a 58-gallon pickle barrel along with water and EM-1. I also added a few scoops of biochar in the bottom of the barrel and some garden soil. The biochar will absorb nutrients and microbes, and perhaps some toxic gases like hydrogen sulfide. It may also provide habitat for some kinds of microbes. I don't really know what the effect of adding biochar to liquid ferments is. I would not add too much for fear of inhibiting the fermenting bacteria, but at the level of about 1 to 2 percent (similar to what you would add in a feed ration for animals), it should be not detrimental and possibly helpful.

Fermented Foods and Medicines

I have tried adding powdered biochar to sauerkraut and naturally fermented pickles. It does not improve the pickling process at all, and may even hinder it. You do not want to inhibit the pH drop from a lactic acid fermentation. You need the acidity to knock back all the spoiler microbes that are just waiting around ready to attack your carefully preserved food. However, I have seen that some kimchi makers use a few sticks of charcoal in a large pickling crock. A solid stick of charcoal may help to absorb some off-flavors without affecting the pH.

Not all fermentations use *Lactobacillus*. Fermentation guru Sandor Katz gives instructions for making *natto miso* in his book *The Art of Fermentation*.[2] Importantly, he notes that natto is produced by the alkaline-forming bacteria *Bacillus subtilis*. Biochar could be helpful in this case. It's something worth exploring because I would like to make natto for its vaunted health benefits. I would also like to be able to produce *Bacillus subtilis* to use as a fungicide. I use a commercial product for now to beat back alternaria fungus on my tomatoes. It helps, but it is very expensive. Researchers in Pakistan have tested biochar and *Bacillus subtilis* for protecting tomatoes against alternaria and have found that indeed there are synergistic effects from the combination.[3]

More Biochar Garden Recipes

Bokashi is a real workhorse in my garden, and it pairs beautifully with biochar. But there are other ways to use biochar in the garden. I have shared my ideas for using biochar in manure management and composting in previous chapters, but here are some ways to use biochar to solve specific garden problems.

VERMICOMPOST

Worms are fine with biochar as long as it is properly prepared. They are sensitive creatures and some things are appreciated in small doses that in high doses will cause them to flee. I have found that they like the biochar I make (they might

Figure 11.3. Sprinkle a scoop or two of moist biochar on the surface of the worm bin on a regular basis and your worms will thank you.

not like some other kinds of biochar) as long as it is plenty moist. Dry biochar scratches their tender hides. Periodically, I sprinkle a scoop or two of moist biochar on the surface of the worm bin. Worm bins can get a bit acidic from all the food waste. The small white worms called "pot worms" are an indication of acidity. The alkaline biochar helps to balance the bin.

When I harvest the vermicompost, it has bits of char in it that are charged up with nutrients and microbes from the worms. This is a great addition to the garden or to a potting soil mix. I call it "vermichar"—it is worm castings with biochar, not charred worms!

CHAR AND GREENS

"Char & greens" is a way to make a quick, hot compost by mixing fresh biochar with fresh greens. In conventional aerobic composting, you mix greens that are high in nitrogen with browns that are high in carbon to get the right C:N ratio. We have found that we can get the same result with greens and blacks. Take fresh grass clippings or weeds you have pulled from the garden and mix or layer it with fresh biochar at a ratio of about 1:1. The amounts are not that critical, so you will probably get it right. Thanks to Umpqua Biochar Education Team's Scott McKain for this recipe.

BIOCHAR LASAGNA

Lasagna bed gardening is a way of building up soil by layering compost ingredients in thin layers and letting it compost in place. Biochar can be a key ingredient in this process that will help retain more nutrients that could otherwise be lost through rain leaching or ammonia volatilization. You can start with a thick layer of cardboard to suppress turf and weeds. Follow that with alternating layers of browns, greens, and blacks. Some popular ingredients for browns include fallen tree leaves, newsprint, wood chips, and straw. Greens include grass clippings and weeds but also high-nitrogen materials like manure and coffee grounds. Sprinkle the biochar on top of the greens for best effect. Make all the layers fairly thin so the ingredients can interact, and repeat the sequence until you have filled your bed. Top it off with some straw or leaf mulch to hold in moisture. Build yourself a biochar lasagna garden bed and start gardening in no time!

FERTILIZERS

I use an organic fertilizer mix as pre-plant fertilizer in all my garden beds. I apply about one quart of this for every square meter of soil and add a handful into planting holes for transplants. I dribble a small line of it in a row where I am planting seeds. The ingredients change from time to time, but here is a typical recipe that I use:

3 parts seed meal or feather meal
1 part crustacean meal
1 part kelp meal
1 part glacial rock dust
1 part gypsum
1 part vermichar, biochar compost, or biochar/bokashi mix

My recipe is based on the "Complete Organic Fertilizer" recipe found in the book *Gardening When It Counts* by Steve Solomon.[4] You can vary this recipe as you need for your soils. For instance, in the spring when I want some extra nitrogen, I will add bloodmeal. If I need more potassium, I will use alfalfa meal. I use gypsum rather than dolomite lime because my soils are high in magnesium and low in calcium. The sulfur in gypsum is beneficial, too. Rock dust is really important for me because my raised beds are high in organic matter but fairly low in minerals and clay.

PEST AND DISEASE CONTROL

Crustacean meal provides nitrogen and calcium, and it also promotes chitinase, an enzyme that degrades insect shells. I have a problem with symphylans. Symphylans are a native forest soil dweller here in the Pacific Northwest that usually feed on fungal hyphae, but when they migrate into a garden where fungal hyphae are more scarce, they will switch to chewing on fine roots. The result is stunted plants that are starved for nutrients. Luckily, I found that chitinase really helps to control them. Like all arthropods, symphylans have an exoskeleton made of chitin.

I also add a mycorrhizal inoculant in the seed rows and transplant holes. Mycorrhizae are soil fungi that surround the root tips of plants and exchange phosphorus and other nutrients for sugars that the plant provides to them through root exudates. I think that mycorrhizae also help protect the roots from symphylans. I use a product called Big Foot Mycorrhizae that includes biochar as an ingredient. Biochar is a good carrier for microbial inoculants, so I feel confident that this product has plenty of spores to inoculate my plants.

Figure 11.4. Symphylans under the microscope. These pests nearly wiped out my garden one year. Biochar alone did not stop them, but combined with mycorrhizal fungi and crustacean meal, the pests were dealt with.

Symphylans are the only soil-dwelling pest I have had to deal with other than root maggots that come from flies laying eggs around the base of carrots and turnips. Biochar doesn't do much to control those or other flying insects. Wood vinegar does help, though.

Since I have gotten all my soil topped up with biochar, I have zero problems with soil fungal diseases like *Fusarium* and *Phytophthora* (an oomycete). I rarely get any scab on my potatoes. Soil pathogens are pretty much history. Researchers at the University of Naples in Italy conducted a comprehensive review of many studies on biochar and disease suppression. They found that biochar "showed suppression efficiencies of 86 percent for fungi, 100 percent for oomycetes, 100 percent for viruses, 96 percent for bacteria, and 50 percent for nematodes."[5] This is big news, and for farmers, biochar's ability to suppress soilborne diseases might be the most economically valuable aspect of biochar.

BIOCHAR POTTING MIX

Potting mix is expensive. Some of the ingredients in it are not sustainable: Peat is mined from peat bogs that take thousands of years to form, so it is like using fossil fuels—as the peat breaks down, carbon that was stored is now released back into the atmosphere. Vermiculite and perlite are mined from the ground, using heavy machinery. Vermiculite can contain asbestos. Perlite is mined and then cooked at high temperature to make it "pop" so it has more pores to hold water. Coco coir is sold as an alternative to peat, but if you live in North America, it is shipped a long way to get to you. Biochar takes the place of perlite, peat, and vermiculite by providing aeration, cation exchange capacity, humus, and water-holding capacity.

Here is my recipe for potting soil that uses locally sourced, sustainable ingredients. This varies somewhat depending on what I have available, and a lot of similar combinations will work just fine. This is a great area for experimentation.

Figure 11.5. My Christmas cactus after re-potting with biochar. Before, it was a pathetic, scraggly-looking thing.

3 parts washed, composted dairy fiber
2 parts soil from my garden or garden compost
1 part biochar/bokashi mix (half and half dry bokashi and biochar)

Test the biochar/bokashi mix for pH and add more biochar if it is too acidic. If I am planting perennials in pots, I will add some bigger chunks of biochar in the bottom of the pot to "sweeten" the soil down where it can get anaerobic and dank. I repotted my Christmas cactus a few years ago with biochar, and in the first year it more than doubled in size!

Trees and Perennials

Many people are now using biochar when planting trees, vines, and other perennials. I like to add chunks of biochar in the bottom of planting holes for drainage, just like I do in potted perennials. Bokashi is a helpful additive, too. Biochar can be especially helpful when replanting an orchard or vineyard that is susceptible to replant disease, which is a condition where soils build up pathogens and toxic chemicals after the continuous cultivation of the same or related species on the same soil. Apple and stone fruit trees are especially vulnerable, and several studies have found that adding biochar to the planting hole can help young apple and peach tree seedlings survive and thrive.[6]

Michael Phillips, author of *The Holistic Orchard*,[7] describes a method he uses to apply biochar in orchards that combines biochar and hugelkultur:

> I piled brush and pulled root systems (of hardwood trees located in intended apple tree rows) on a downhill slope and deliberately scorched the branch ends and larger stump surfaces with a dry leaf fire. Notch one up for crude biochar made in place. Uphill soil was then pushed over the works to create a level planting plateau. . . . The apple trees planted at the intersection of this altered slope are the healthiest on my farm.

In Japan, arborists have used charcoal to revive ailing old trees that grace the grounds of sacred shrines and temples. Arborists carve shallow trenches in a radial pattern from the tree's trunk to the edge of the dripline and apply fine-grained charcoal inoculated with mycorrhizal fungi. This is successful in reviving these well-loved trees that have suffered soil compaction from their many visitors. A similar project is underway in Britain, where U.K. biochar company Carbon Gold has partnered with Apex Soil Solutions to use their soil injection equipment to inject Carbon Gold's Biochar Tree Soil Improver product into the root zone of Bexley's Charter Oak, a 200-year-old oak in London's Danson Park. After several years, the results have been spectacular. According to Carbon Gold: "In 2022 we went back again, and . . . this majestic tree looked simply beautiful and was as healthy as ever—robustly rooting itself into history for many more years ahead!"[8]

Biochar Mulch

Biochar is great as a mulch around the dripline of trees and other perennials. I used a thick mulch of plain biochar around my raspberries and it does a good job of suppressing grass and weeds and holding available water for the raspberry roots. Raspberries need abundant water during flowering and fruiting, but they don't like waterlogged soil. Biochar has the unique ability to hold air and water at the same time, so it holds water in soil without becoming waterlogged and anaerobic. How does that work? Water is attracted to the charged surfaces inside biochar pores, while at the same time, the larger biochar pores are filled with air. I do a periodic soil drench with EM-1 and fish hydrolysate to keep the pH low and feed the berries. My neighbors tell me that they have to plant new

Figure 11.6. A heavy mulch of plain biochar has been suppressing weeds and holding water in my raspberry beds.

raspberry canes every five years or so because the old ones peter out. My canes have been producing abundantly for more than fifteen years now.

Tips for Applying Biochar to Soil

There are many ways to use biochar in soil. All of the recipes here for composts, ferments, fertilizers, and growing media are vectors for adding biochar to soil. If you are already using compost and dry fertilizer mixes, apply biochar just by adding it to these amendments. If using spreading or seeding equipment, you may need to crush or grind the biochar to a uniform size so it flows through the equipment. If you have limited amounts of biochar, band it in the seed row with fertilizer where it will be immediately accessible to growing roots.

You can add plain biochar without inoculation, as long as you don't add too much at once. Spread about ½ inch (1 cm) of biochar on the surface of the soil and work it into the top 4 inches (10 cm). If your soil is already healthy, microbes will colonize the biochar quickly. If you add it in the fall, the soil will have more time to adapt to the biochar amendment before your crop starts growing. The best time to apply plain, uninoculated biochar is in the fall with a legume cover crop, because the legumes will fix nitrogen.

Recipes for Large Farms

My direct experience with biochar in soil is limited to gardens, and gardening is definitely not the same thing as farming. One of the biochar pioneers in the United States is Phil Blom, owner of BioEnergy Innovations Global Inc. Phil has been distributing his biochar brand, TerraChar, to commercial farmers for many years now, so I asked him what kind of help and advice he provides to his customers. I was surprised to learn that he tells farmers to take the same approach that I recommend: Use biochar to manage manure and make great compost. Then apply compost as usual. Phil explained that the farmers who are most likely to see the value of biochar are those who already understand the value of manure. The farmers he works with often own large poultry, beef, swine, or dairy operations and grow organic crops on five hundred acres or so. Most of these farmers are Amish or Mennonite. They are more grounded in reality, Phil says, because they know how important it is to recycle nutrients. Here is the recipe one farmer is using for on-farm compost with poultry waste.

Ingredients

5 semi loads of TerraChar biochar (about 55 cubic yards [42 m^3] a load)
2 semi loads of dead birds from the egg-laying barns
5 semi loads of poultry litter cleaned out of the poultry barns

Procedure

1. Lay 16 inches (40 cm) of biochar to make a pad.
2. Add the dead birds.
3. Layer half a load of biochar.
4. Layer half a load of poultry litter.
5. Top with a half a load of biochar to seal the mound.

Let this sit for three weeks while the dead birds decompose. Then mix in the rest of the biochar and poultry litter and let it compost for ninety days, turning it after each thirty days. This makes a rich fertilizer with blood, bones, feathers, and manure, stabilized with biochar. Biochar controls the odors throughout the process. At the end it just smells like highly organic fertile soil.

Blom has tried to interest conventional farmers in using biochar, but there are often too many barriers in the way, even though the farmers who have tried it have had great success. Phil Blom and David Yarrow helped Missouri farmer J.R. Bollinger develop a system for using biochar on his five thousand acres, where he mostly grows corn. Blom and Yarrow developed formulations that J.R. could use in his no-till system that include a light pre-plant application in a strip tilling pass, followed by more biochar added at seeding. J.R. also includes a fine biochar powder in a fertigation step. Some equipment modifications were needed to make the applications flow, but J.R. was up to the task.[9]

J.R. Bollinger achieved record corn yields in the state of Missouri, but not many of his neighbors have followed suit. Sadly, the farming system in the United States is almost completely controlled by big ag, chemical, seed, and fertilizer companies.

Suppliers like TerraChar can provide biochar to farmers in bulk semi loads or bagged in 2-cubic-yard (1.5 m^3) totes. To find biochar suppliers in North America, see the Producer Directory on the USBI website, biochar-us.org.

Biochar Extension and Support

Fortunately, the biochar industry is making great progress in developing and providing good extension information about biochar. Academics have produced a huge body of research on biochar in agriculture, so there are abundant knowledge resources available for study. Biochar industry groups in several countries and regions offer fact sheets, reports, webinars, and videos that contain accurate information on biochar. Many state university extension departments in the United States and Canada offer biochar information programs tailored for regional crops and producers. When I was on the board of the US Biochar Initiative, I started the Biochar Learning Center, a searchable online database of biochar information provided by many of these different sources. This is a good place to start if you have specific questions about your biochar application. You can find the database at biochar-us.org/learning.

The US Department of Agriculture has several programs coming online to develop and promote the use of biochar in agriculture. The USDA Agricultural Research Service (ARS) funds biochar research programs at institutions around the United States. Researcher Kristin Trippe's USDA-ARS lab at Oregon State University is developing an online tool called the Biochar Atlas to match biochar properties with soil needs. This tool is integrated with the online USDA Web Soil Survey that provides soil details for most of the United States. Producers can input their locations and find guidance on biochar based on their soil type. Because soil type maps don't include information on farmer soil inputs, the tool allows a grower to enter information from soil tests as well. The tool makes recommendations based on soil nutrients, pH, carbon content, and soil moisture. For instance, a farmer who grows wheat on acidic soil that needs liming might find a recommendation for an alkaline biochar with a higher ash content. Dr. Trippe's group continues to build the functionality of this tool, and soon it will include recommendations and economic cost-benefit analysis for many more crops. They also plan to integrate it with the USDA soil carbon and climate monitoring program called COMET-Farm.

In addition, the USDA Natural Resources Conservation Service (NRCS) is rolling out a new cost-share program to help farmers afford biochar applications. The NRCS Soil Carbon Amendment (Code 336) protocol is now available through the NRCS Environmental Quality Incentives Program (EQIP). The Biochar Atlas will be an important tool to guide farmers and crop advisors on how to specify and apply biochar under the Code 336 protocol. If you are in the United States and you are interested in taking advantage of this program, contact your local NRCS office for more information. You can also find a series of fact sheets and guidelines at the USBI website, biochar-us.org.

CHAPTER TWELVE

For the Forest

If you have a piece of land with woody debris on it, you can easily make your own biochar to use on your farm, garden, or livestock operation. Biochar can solve many problems for those who live on the land. But biochar also has a role to play in solving the big environmental problems that threaten all of us: deforestation, drought, pollution, climate change, and more. Saving forests, with all of their biological diversity and ecosystem services, is at the top of the list.

In the western United States, drought and a century of industrial forestry and fire suppression have created a wildfire crisis that threatens ecosystems and communities. As forests go up in smoke, we are also experiencing the loss of one of our most important natural carbon sinks, at a time when we must rely more and more on natural climate solutions to draw down carbon. Increasingly,

Figure 12.1. The Illinois Valley Fire Resiliency Oversight Group (IV-FROG) on a field trip to look at fuel reduction needs in our overgrown forests.

scientists and land managers are investigating human actions that can help restore these forests and their ecosystem services. The manual thinning and removal of excess biomass is one of the most important actions we need to take to improve forest resilience.

The scale of the problem of excess fuels in forest landscapes is massive, and implementing thinning programs in areas where they can help control wildfires is costly and requires a significant mobilization of labor and equipment. Furthermore, there are few alternatives for disposing of the resulting slash materials.

Some of the material is chipped and transported to biomass energy plants, but that is only feasible when the distances are short (less than 40 to 60 miles [64–97 km]). Some material is masticated and left in place. A masticator is a machine, usually mounted on a tractor, that moves through a forest and chews up the slash and spits it out on the forest floor as a mulch. Mastication is limited to gentle slopes where the machinery can operate. The potential for soil compaction exists on some soil types and conditions, and if the mulch is greater than 4 inches (10 cm) thick, it can suppress regeneration of native species. Thick layers of mulch can also become a fire hazard that once ignited can smolder for a long period of time. Mastication can be beneficial on the right site and if done with care, but it is a costly operation.

The most economical alternative for biomass disposal is the current practice of incinerating it in on-site burn piles. Slash pile burning may alter soil microbial populations, destroy seeds, and result in bare soil, which is vulnerable to colonization by invasive species.[1] Smoke from slash pile burning limits the burning window especially in air-quality-limited watersheds, making it more difficult to accomplish the work.

As we have seen in chapter 3, it is perfectly possible to make biochar in the woods and leave it there as a viable alternative to burning piles to ash. Biochar offers many benefits to forest ecosystems. Returning biochar (and natural, low-intensity fire) to natural ecosystems should be our first priority, even before agriculture. Why? Because it is forests and other ecosystems that provide the bulk of the biomass feedstocks we will need to use to recarbonize agricultural soils. This is like asking why you should put your own oxygen mask on first in an in-flight emergency, even before helping children: because otherwise you will run out of oxygen and be unable to help anyone else. If we don't have healthy forests and other ecosystems, we not only diminish their "net primary productivity" (how much biomass they can grow), we also lose their many ecosystem services (like clean water), and we lose their function as carbon sinks for climate mitigation.

Making biochar in the woods has now become a movement that aims to mobilize a workforce to accomplish the task of putting the carbon back into forest soils. "Biochar in the Woods" is a subset of a larger idea called place-based

Figure 12.2. On the top, an incineration burn pile has destroyed the organic layer leaving bare mineral soil vulnerable to erosion and weeds. The photo on the bottom was taken one year after a conservation burn that left the organic soil intact with a layer of biochar above it.

biochar, or the practice of making biochar in place and using it right there, whether it is a forest, a farm, or a city park.[2]

Years ago, in my tree-hugger days, I protested the logging of old-growth trees. We held vigils and roadside protests at sawmills in the Pacific Northwest and we would often see trucks go by that carried huge logs that were from 500-year-old trees. Some loads might have only three or four of these giant logs. There were even "one log loads," a log truck loaded with just one 10-foot (3 m) diameter log. We would stand there on the side of the road and yell, "Put 'em back!"

Figure 12.3. Leaving biochar in the forest improves forest soils for herbs, orchids, fungi, and trees.

Well, we cannot put back the old-growth forests that once carpeted North America, but we can put the carbon back in soil and set the stage for new growth and regeneration. Before discussing how we can mobilize to accomplish this task, we should understand more about what biochar can do for forests.

Functions of Biochar in Forest Soils

Biochar is a useful addition to our agricultural soils, but it also has a critical role to play in the ecological restoration of forests, grasslands, wetlands, and other landscapes. In particular, fire-adapted forests really need to get their biochar back. People can make biochar, but Nature makes biochar all on its own. Charcoal from historic wildfires is an important component of soil that has been depleted in forests where fire has been excluded, and we need to put it back.

Soil charcoal is a product of natural and anthropogenic vegetation fires that take place globally on many scales. Nature makes megatonnes (an estimated 256 megatonnes per year) of black carbon during wildfires and prescribed fires in forests, and that equates to 12 percent of the total carbon that is emitted by landscape fires across the globe every year.[3] A global inventory estimates that on average 13.7 percent of the total organic carbon found in soils is pyrogenic carbon, or biochar. Some soils have up to 60 percent of their organic carbon in the form of biochar.[4]

Figure 12.4. After the 2020 Slater Fire in southern Oregon, char is found everywhere, including deep in the soil where fire has charred a tree's root system.

The amount of charcoal generated by wildfire depends on fire intensity, fire return interval, vegetation type, fuel loading, and fire behavior. Researchers from the Wilderness Society found that the conversion rate of biomass to charcoal during a forest fire event ranges from 1 to 10 percent of the biomass consumed in a fire.[5] Based on biomass inventories, they estimated that a single fire event in a mature lodgepole pine forest might deposit 3.25 tonnes per hectare of carbon in the form of charcoal. They concluded: "Thus, wildland fire need not be viewed only as a cause of [carbon] loss to the atmosphere, demanding suppression, but rather, as a driver of long-term [carbon] sequestration."

In some forest types and under some conditions, conversion rates of woody biomass to charcoal are higher. A postfire inventory of an experimental high-intensity crown fire in a Canadian boreal forest stand found that 27.6 percent of the carbon in the fire zone was converted to charcoal. Extrapolated globally, the researchers speculate that charcoal production in boreal forests could be as much as 100 megatonnes annually, more than five times the previous estimate.[6] North American boreal forests have soil charcoal levels that are two to three times higher than in Eurasian boreal ecosystems, where fires are less intensive.[7]

When considering the impact of biochar on soils and plants, it is important to recognize that biochar materials are variable and may produce different results depending on soil type and other factors. Charcoal made in a forest fire is much more variable than industrially produced biochar, or even biochar made in a flame-cap kiln, containing everything from torrefied (roasted) wood to mineral ash, depending on the spatial distribution of fuels and fire intensity. Here is a summary of the effects of biochar in forests.

Nutrient Cycling. Biochar can have a strong impact on soil nutrients and their availability to plants. Both biochar and ash contain some nutrients. Due to its sorption properties, biochar can bind and retain certain nutrients for short or long periods of time. Biochar also affects the soil biological community and its role in nutrient cycling. Since fire is the dominant form of disturbance in western forest ecosystems, the exclusion of fire is likely having a significant impact on forest nutrient cycles.

Ash. The fertilizing effects of ash following fire are well known. Wildfire charcoal contains mineral ash that is a source of soluble nutrients such as potassium, phosphorus, calcium, and magnesium.

Liming, pH, and CEC. Char and ash have impacts on soil pH, base saturation, and cation exchange capacity (CEC). All of these properties are involved in nutrient cycling. Biochar can improve soil CEC by providing numerous negatively charged cation exchange sites on the biochar aromatic carbon matrix. The combination of ash and char is likely to be more effective than lime alone or ash alone in promoting soil nutrient cycling.[8]

Nitrogen. Nitrogen volatilizes in a fire, but ammonium, a combustion product, is often left behind, especially in fires that burn less severely at lower temperatures. Biochar is especially effective at sorbing ammonium. Depending on conditions, char sorption of ammonium can work like a carbon-based slow-release fertilizer.[9]

Nitrification. Nitrification is the biological process that converts ammonium to nitrates that are easily used by plants; biochar enhances nitrification in forest soils.[10] This effect may be especially important in forest soils where nitrogen is limited. The charcoal effect is due to a combination of factors:

sorption of phenolic compounds in soil that inhibit nitrifier bacteria, and other properties of biochar that seem to promote the microbial community of nitrifiers.

Mycorrhizal Fungi. Biochar is reported to change the abundance and species distribution of fungi, bacteria, and other soil life-forms.[11] Biochar can have an especially beneficial effect on the mycorrhizal fungi that are essential for healthy forests. A study of charcoal used with larch tree seedlings found that biochar stimulated growth of roots and mycorrhizal fungi.[12] In this study, the fungi solubilized soil phosphorus that was otherwise unavailable to the seedlings. The phosphorus then showed up in greater concentrations in the needles of the seedlings grown in biochar-amended soil.

Soil Water-Holding Capacity. One of the most important effects of biochar is its impact on soil water-holding capacity and hydraulic conductivity (the movement of water through soil). Because biochar is porous, it can absorb water like a sponge. Water is held in biochar pore spaces and voids, and in the spaces between particles in the soil. Biochar interacts with other soil constituents to form aggregates, which also function to retain water in soil.[13] The impact of biochar on soil water-holding capacity has a lot to do with soil texture. A series of experiments in the 1940s looked at the impact of charcoal on three different forest soil types based on sand, clay, and loam. Using different percentages of charcoal addition, different types of charcoal (hardwood and softwood), and different particle sizes, the results showed that, overall, charcoal greatly increased the water-holding capacity of sand, slightly increased the capacity of loam, and improved the drainage of clay soil.[14]

Impact on Tree Growth. Most biochar plant-growth studies have been performed on field and horticultural crops. However, there are a growing number of studies on biochar and forest species. Many of these have been pot studies of forest tree seedlings. A meta-analysis has summarized a number of these studies on responses of woody plants to biochar.[15] The analysis found a significant tree-growth response to biochar, with an average 41 percent increase in biomass. This is a highly significant result, but it is important to understand that since many of the studies were of tree saplings in early growth stages, it is unlikely that this kind of response would apply to older trees with much slower growth rates. However, if biochar improves the early growth of tree seedlings, it would support the idea that biochar is a valuable tool for reforestation.

Trees in my backyard have demonstrated their love for biochar. I had a tote bag half full of biochar that I left in the yard for about six months, all through the summer dry season. In the fall, I wanted to shovel some of the char into a

Figure 12.5. Pine roots invaded a bag of biochar in my yard one year.

wheelbarrow to use. I found the char pile completely riddled with tree roots from a nearby stand of pines. The roots had penetrated through the bottom of the plastic tote bag and were covered with char particles, held by the intrusion of fine root hairs into the biochar pores. It's hard for me to figure out what they were after. The biochar was a gift from Oregon Biochar Solutions. Their Rogue Biochar is shipped with 30 percent moisture for dust control. But this bag had sat out in the heat with no rain for the whole summer. It should have been very dry. Were they after the minerals from the ash? Were they looking for free electrons from biochar surfaces? Or was there enough residual water held by electrostatic attraction to pore surfaces to attract this kind of root growth? It was very curious and a reminder that there are many biochar mysteries yet to solve.

Managing Forests for Biochar

Given the widespread presence and many functions of charcoal in fire-adapted forest soils, we need to recognize the impact of fire exclusion on forest-soil charcoal and consider management changes that could restore it. A few studies have attempted to correlate soil charcoal with the history of fire suppression in order to estimate the charcoal deficit in forest soils. Looking at different site histories in ponderosa pine / Douglas fir forests of the inland northwest, researchers found that sites that experienced multiple fires during the past 79 to 130 years contained about three times more charcoal than forests where fire was excluded.[16] A modeling exercise that compared different fire return intervals in a 100-year-old ponderosa pine forest found that more frequent fire is likely to promote the accumulation of char in the mineral soil layer and slightly more soil charcoal overall, while less frequent, more severe fire left more charcoal in the soil organic horizon.[17]

Changes in the fire regime could potentially restore natural biochar to forest soils, but there are still many site-specific considerations and unknowns. One of the goals of forest restoration in western North America is to re-establish the natural fire return interval through a combination of strategic thinning and controlled burning, eventually allowing fires to burn more naturally and with less severity across larger landscapes. An examination of soil charcoal in a forested landscape in the Pacific Northwest found that soil charcoal was correlated with site microclimate conditions such as temperature and moisture.[18] Warm-dry sites produced less charcoal than cool-wet sites, and less intense fires may produce more charcoal.[19] On the other hand, frequent fires can potentially incinerate char that remains in the litter layer, reducing the accumulation of charcoal, but that is not always the case.[20]

All of these site condition factors should be considered along with burning methods when determining fire prescriptions for optimizing soil charcoal formation. For instance, during a prescribed fire in north-central Florida, researchers tested char formation under two different fire-spread patterns: a head fire (with the wind) and a backing fire (against the wind). They found that backing fires formed more than twice as much charcoal as head fires due to differences in oxygen availability and residence time.[21] Backing fires move forward against the wind, with the flames bending over the already burned area behind the flame zone. Backing fires produce fewer emissions because more fuel burns in the efficient flaming phase of combustion than in the smoldering phase. Backing fire conditions can also be created by burning from the top of a ridgeline down, as wind and draft tend to move uphill.

Traditional Ecological Knowledge

How do we distinguish between "natural" fires and human-caused fires? The fact is that anthropogenic fires have been a factor on planet Earth since humans first mastered fire, 100,000 years ago or more. Sometimes these fires seem to benefit ecosystems and sometimes they are clearly detrimental, like the recent trend toward megafires fueled by global warming, drought, poor forest management, and careless human ignitions. Historically, humans have used fire extensively to alter environments and make them more productive by cycling nutrients. This productivity can benefit a multitude of species, not just humans, and this was by design.

Traditional, Indigenous dwellers on the land were place-based people and they understood that Nature would provide for them if they respected the whole of life. Setting frequent landscape fires at the right place and time was part of living on the land. Living on the land in this way makes humans a part of, and inseparable from, Nature.

David West, who was the Native American Studies director at Southern Oregon University until his retirement in 2015, is a citizen of the Potawatomi Nation. David has joined us for several of the biochar camps held by the Table Rock Foundation in Josephine County, Oregon, for youth and adults, where we practice making biochar in stoves, kilns, piles, and landscape burns. We have enjoyed hearing his stories about traditional ways, and one of the stories that has really stuck with me is the one about how the women burned the forest from the top down.

At the right time of the year, women would travel to the mountain ridgetops and light fires that would slowly burn down through the high mountain meadows into the forest below. These were like the backing fires (against the wind) that prescribed fire technicians use to create moderate burning conditions. As the fires crept down the mountainside, they would tend to burn the understory and not leap into the canopy to cause a general conflagration. Of course, this would happen at times, but the patches of severely burned forest contributed to biological diversity by providing early seral ecosystems with burnt snags, shrubs, and brush for species that favor those conditions, like woodpeckers and songbirds. Other early seral species such as manzanita, hazel, and bear grass were important food and fiber species for people. At regular fire intervals of three to twenty years, the understory stayed clear, breaking the fuel ladders that let fire climb into the tree canopy.

In a paper titled "The Importance of Indigenous Cultural Burning in Forested Regions of the Pacific West, USA," authors Jonathan Long, Frank Lake, and Ron Goode survey the growing literature on what is known as "cultural burning" and its relevance to our current predicament in the age of megafires.[22] Traditional fire management included backing fires to burn out firebreaks to keep a wildfire from engulfing a village. They kept ridgelines cleared with fire breaks that also enhanced travel corridors in the mountains, and lit fires on a regular basis to clear fuels from areas around villages, trails, and gathering sites. The authors explain that this approach was "not intended to stop fires with stark fuel breaks, but rather to facilitate control of fires and to moderate their intensity and severity." They note similarities to the modern fire control strategy of developing PODs (potentially operational delineations), which are a series of strategic firebreaks that the US Forest Service and other agencies have begun to implement.

They also note that native people in the Pacific Northwest understood that incorporating char into soil benefitted plant growth. Special herbs like tobacco were sown into burned areas where char, ash, and condensed smoke would aid germination and growth. Smoke that cooled and condensed on the vegetation would protect important stands of willow and oak from insect pests. The people also noticed that animals would eat char to help rid themselves of parasites.

Miraculously, considering the violent disruption of their traditional ways, tribal members still retain stories and memories of cultural burning practices that they continued even surreptitiously up to the present, and they are willing to share some of these traditions. Cultural burning and modern fire science are beginning to merge into one worldview that recognizes the mistakes of colonizing Europeans in their approach to managing the forests of the West, and charts a path forward to renewal. For instance, the US Forest Service is beginning to collaborate with groups like the Western Klamath Restoration Partnership, which is taking responsibility for managing fire on traditional lands of the Karuk Tribe in Northern California.

This is a dynamic process that is still reaching for shared understanding. Cultural burns are integrated into a place-based culture that includes multiple layered values and close monitoring over generations. Prescribed burning projects by land management agencies have more limited objectives, mostly focused on reducing fire danger. Perhaps this limited focus is somewhat justified for now because there are so many land management mistakes to correct and there is no way to reintroduce better practices without first reducing the dangerous fuel-loading conditions that have been created. The goal of forest-fuel thinning and prescribed burning should be to remove enough fuels so that tribes can restore their traditional cultural burning practices and become self-reliant on their lands once more.

One of my personal heroes is Sean Hendrix, base commander for Grayback Forestry, a wildland firefighting company. Sean is an advocate of prescribed burning, but in recent years the issue of liability for escaped burns has severely limited his opportunities to put good fire on the land. Sean has responded by finding opportunities to light good fires whenever he can during a wildfire incident. In 2017, Sean was part of the incident command for a large fire in Douglas County, Oregon. The other fire bosses wanted to start a burnout fire at the bottom of a steep slope full of old-growth forest to make a wide firebreak to protect roads and houses in the valley. Instead, Sean lit fires at the top of the ridgeline to create the firebreak there. The fires burned part way down to the valley, but the backing fires put themselves out when the upslope draft blew the flames back over the burned area. Sean created the firebreak, saved the houses and the old-growth forest, and put good fire and biochar on the ground. Sean is now a specialist in the technique. To safely start these ridgeline ignitions, he drops ping pong balls loaded with napalm from a helicopter. Sean is reviving the approach of the native women who invented the top-down burning of fire-prone landscapes. This approach saves lives and trees.

Many different collaborations between land managers, prescribed burn associations, and Indigenous fire practitioners are taking place now. It will be very interesting to see how these efforts will evolve to return control of forest fire management to those who are most connected to the land. You can follow the efforts of the Indigenous Peoples Burning Network and other fire networks at firenetworks.org.

Putting the Carbon Back

The US Forest Service Wildfire Crisis Strategy calls for treating up to twenty million acres over ten years on national forest lands in the West. Treatments to remove biomass from forests include various combinations of mechanical thinning, chipping or mastication, and prescribed fire. With millions of acres needing treatment, managers are looking for ways to help pay for logging and residue disposal costs. Some sites can be commercially thinned which often pays for slash pile burning or mastication; however, many sites do not have commercial-sized timber or removal of large trees is not warranted for watershed health and restoration of late-seral conditions (old-growth forest).

The Forest Service has been supportive of flame-cap kilns and conservation burn approaches to burn piles for making biochar. The Forest Service also has a development agreement with Air Burners to produce a mobile air curtain burner called the CharBoss—basically an insulated firebox on wheels with a blower to control the air. This works much like a flame-cap kiln, except that it has a moving grate that conveys finished char out of the bottom of the fire box so the process can operate continuously. The blower keeps any smoke emissions to a minimum and it helps with burning wet fuel. Tigercat manufactures a larger version of this concept. The blower used in these machines makes it possible to quickly consume large logs that would burn very slowly in a passive flame-cap kiln like the Ring of Fire or the Big Box. However, a consequence of this air-assisted incineration is more ash and less biochar.

Life cycle assessment (LCA) is needed to determine the overall carbon footprint of any technology, including the option of processing and transporting biomass to where it can be used for heat and power. The answer will most often be to restrict biomass energy to locations that are close to the woody debris. The Waste to Wisdom project was a comprehensive study of ways to make use of some of the excess woody biomass causing so many problems in our forests now. I worked on the portion that did LCA comparing the alternatives of burn pile incineration to flame-cap kilns to a mobile gasifier.[23] The flame-cap kilns had the lowest carbon footprint because, unlike the gasifier, there was no requirement for diesel fuel to chip, dry, and process the material in the complicated gasifier machine. Flame-cap kilns were superior to pile

Figure 12.6. Alternative fates of forest slash. What are the consequences of the alternative treatments?

burning because they emitted only half of the CO_2 and made biochar for long-term carbon storage.

One of the largest biomass power plants that makes biochar is right here in southern Oregon. In 2020, the Labor Day fires torched thousands and thousands of acres of forest around us. We know those forests will burn again, so the US Forest Service is trying to get ahead of things and is proposing seven thousand acres of fuel removal along roadsides so we have more firebreaks and a better chance of controlling the fires. They tried very hard to get the power plant to take the fuel—for free! But they cannot afford to transport it. It is purely a matter of distance. The power plant has all the fuel they need close by, so why spend the money on faraway fuel? Until we have more biomass facilities distributed on the landscape, the best alternative is to make biochar and use it on-site.

I hope we will get to make biochar, on-site, using flame-cap kilns and conservation burns, from some of the slash that will come out of the project. I will advocate for leaving the biochar on-site to help hold water in forest soils and revegetate ground that has had much of the carbon burned out of it by fire. The alternative is more ash and more smoke and particulates from the uncontrolled pile burns they will use to incinerate the fuels.

The New Carboneers

At the end of the day, the question is: How can twenty million acres of fuels be converted to soil-building biochar? Who will do the work? And who will pay for it?

Traditional ecological knowledge tells us that there should be no question of paying for this restoration work. We owe it to the land as our stewardship obligation. While as a society we have not yet embraced this ethic, as individuals many of us have. Hundreds of farmers, woodland owners, tribes, and land conservancies have begun reducing fuels on their properties and making biochar to use in farms, gardens, livestock barns, and for ecological restoration.

At the same time, a network of Prescribed Burn Associations in multiple states is forming to address fuels and put good fire back on the land. Prescribed Fire Training Exchanges (TREX) are active in networking many fire communities and providing hands-on training on the ground to increase the number of experienced burners. Prescribed fires conducted by TREX-trained burn technicians create more fire-adapted landscapes for the benefit of forests and local communities.

Prescribed fire leaves behind significant amounts of biochar as it burns "low and slow," returning good fire to the land. But many places are too thick with overgrown fuels for "low and slow" fires and must be thinned first. Processing the slash material in biochar kilns or conservation burn piles is another form of "good fire" that we need. Biochar combined with prescribed or cultural burning is the way forward for forest recovery. Advocates of prescribed fire and of biochar are beginning to find each other and reinvent new place-based methods for forest regeneration.

As we find our way forward, there is much to learn, especially about making biochar in the woods at a large scale. In 2021, the US Biochar Initiative held a Biochar in the Woods symposium that brought a dozen practitioners together with officials from federal agencies to share information and lessons learned from our first attempts at making biochar in the woods. In order to keep the conversation going, we started a Biochar in the Woods online discussion forum and a monthly Zoom meeting for brainstorming and discussion. The forum and meetup are open to anyone who wants to know more and share their ideas and experiences. You can find us at biochar.groups.io, a site that hosts several biochar discussion groups. The site is owned and managed by Tom Miles, director of the US Biochar Initiative and one of the persons most responsible for developing and promoting the emerging biochar industry.

I like to call our new place-based biochar workforce the New Carboneers. Historically, the independent entrepreneurs who made charcoal in the European forests had a special place in society. Living on the fringes, they were sometimes seen as clever, wise, and magical people, and other times as lowly drudges. Folklore from medieval times "often attributes to them courage and cunning, the ability to use their force and wit to cope with many dangers, factual as well as symbolic (ranging from philosophic issues to the struggle against savage beasts). . . . It is no coincidence that charcoal burners have become the legendary forefathers of the prominent Bohemian and Moravian noble families. . . ."[24] In France they were called *charbonniers* and in Italy they were known as the *Carbonari*. The Carbonari were active throughout the turbulent political events of the nineteenth century in Italy. The Carbonari were revolutionaries who hatched secret plots in their remote woodland hideouts to replace the absolute monarchy with a constitutional republic. Our New Carboneers are plotting to change the way we relate to forests and fire and bring about a new paradigm of forest regeneration and renewal. Future generations will revere them.

The Carbon Conservation Corps

The place-based biochar production technologies described in chapter 3 are characterized by low capital and operating costs but relatively higher labor costs than most other slash-treatment methods. Labor-intensive methods hold particular relevance to our time, as economic disruptions have resulted in job losses and changes to entire industries. Governments are now confronted with a need for economic recovery at a time when economic growth is slowing, debt is multiplying, employment is falling, and energy costs are on the rise.

Just as the Civilian Conservation Corps (CCC) helped save a generation from poverty during the Great Depression, we propose that a similar program, a "Carbon Conservation Corps," could help our current generation recover from both economic and environmental catastrophe.[25] Producing biochar from biomass that is thinly distributed across sensitive landscapes with high ecological value requires a large workforce for implementation. One benefit of this can be captured by a social life cycle analysis approach, which would identify one key advantage: The money invested in on-site biochar production goes into the pockets of citizens rather than fossil fuel companies for transporting large quantities of biomass. The CCC approach also offers a range of opportunities for skilled labor, from machine operators and arborists to firefighters and ecosystem-restoration technicians.

Organizing young workers into cadres for field work has other benefits besides employment, skill-building, and wages. It helps young people develop self-discipline and a work ethic. Furthermore, in a time of social division, stress, and uncertainty, the experience of working together in a team can help promote social cohesion and even lifelong friendships and connections. Commentators observed that the success of American soldiers in prevailing in the battles of World War II likely owed a great deal to "Roosevelt's Tree Army," the CCC. During the nine years of the CCC's existence, approximately five percent of the total male population of the United States was enrolled in the program. This trained, disciplined workforce provided "pretrained manpower to fill the U.S. Army's ranks upon mobilization."[26]

The potential of biochar for ecosystem restoration is inspiring more and more organizations to take on the task of organizing crews to do the work. Here are a few stories of projects that have begun. There are many more that are getting started now.

Figure 12.7. California Conservation Corps crew members make biochar in the Ring of Fire biochar kiln in the Usal Redwood Forest. A water tank on a trailer with an electric pump stands by to quench the biochar when the kiln is full. Photo courtesy of Redwood Forest Foundation, Inc.

Usal Redwood Forest Biochar Project

Recently, the Redwood Forest Foundation (RFFI), a private conservation group, initiated a project to convert forest slash from thinning operations into biochar. RFFI owns the Usal Redwood Forest, where they are engaged in many ecosystem-restoration activities and biochar projects in partnership with the Sonoma Ecology Center. RFFI is creating fuel breaks and converting the removed slash into biochar in the forest. The biochar is being placed in the forest soil to improve nutrient and water retention, while sequestering carbon. During the winter of 2020 to 2021, they employed a California Conservation Corps crew to thin overcrowded young stands of Douglas fir and tan oak to improve health and fire resiliency of the remaining trees. They completed a total of four days of biochar production as part of the work.

Karen Youngblood, restoration forester for RFFI, reported that after cutting to length and staging the material on a landing, the crew operated Ring of Fire and Oregon kilns for four days, producing 48 cubic yards (37 m^3) of biochar from approximately 336 cubic yards (257 m^3) of slash. Karen has also found that the Ring of Fire kiln works very well for disposing of mountains of invasive broom that the California Conservation Corps crew cut and pulled out of the forest. She says that broom mixed with some tree debris burns well and makes a very nice biochar. Broom is a nitrogen-fixing legume so it would be interesting to see if the broom biochar has a high nitrogen content.

Islands Conservation Corps

The San Juan Islands Conservation District launched the Islands Conservation Corps (ICC) in 2020 to create and train an ecology workforce. The program addresses several conservation objectives including oak woodland restoration and community wildfire protection. Some support comes from the electric utility to clear vegetation along utility corridors.

The crew is recruited from students at Western Washington University's College of the Environment, Huxley College. Program manager Kai Hoffman Krull said: "We had seventy applicants for the program and chose only ten. The students get a stipend and twenty free college credits. With this program, they are gaining many valuable skills in chainsaw work, burning practices, planting, riparian zone restoration, all in coordination with their academic coursework. We are bridging that gap between the university and the field."

The work is done using the conservation burn technique. They make standard-size burn piles, about 5 feet (1.5 m) in diameter. They are lit on the top to get a good flaming combustion going that reduces smoke. They have a 1,300-gallon water truck for quenching and use a combination of spraying water and raking. For very remote sites, they have 5-gallon backpack pumps. One backpack pump can extinguish two piles.

Fire Ecology Restoration Project and the Yew Creek Project

The Fire Ecology Restoration Project (FERP), part of the Permaculture Institute of North America (PINA), is offering paid work, internships, and research opportunities to those who are passionate about returning managed, low-intensity fire to western forests, especially those who are seeking a career path in forest restoration or who have land where they would like to implement state-of-the-art restoration forestry. Making biochar from slash piles is one of the learning objectives of the project.

The FERP program is providing training from experienced carboneers Tomi Hazel Vaarde, steward of Wolf Gulch and author of *Social Forestry*,[27] and Ken Carloni, biology professor and manager of the Yew Creek Land Alliance. Carloni and his Yew Creek partners were pioneers who were the first to design and implement a biochar forestry project that received funding from the NRCS, under a special program that supports making biochar in the woods.[28] They continue to innovate with new kiln designs and approaches to forest management and to monitoring the impacts of biochar in the forest.

Howard EcoWorks

Based in Maryland, Howard EcoWorks is an ecological restoration organization with a mission to empower communities and develop skilled and diverse workforces to restore natural ecosystems. The nonprofit works with state, county, and private landowners to restore degraded forests, riparian areas, and other natural ecosystems in Howard County. Much of the work involves removing invasive shrubs and trees like privet, paulownia, and autumn olive. They have begun using biochar kilns (Ring of Fire and Oregon kilns) to convert this unwanted biomass to biochar. The biochar is used in pollinator gardens with native plants. Biochar added to soil can improve water infiltration and reduce runoff during storm events. Howard EcoWorks is working in partnership with the University of Delaware to establish test plots with biochar for stormwater management.

Howard EcoWorks director Lori Lilly is enthusiastic about the role of biochar in restoring a healthy environment in the region, and also its ability to offset the carbon footprint of the restoration activities. The work is funded partially through grants, but one third of its funding comes from landowners paying fees for the services that the work crews provide. Lori would like to do more biochar projects for landowners but finds that there is limited knowledge about the benefits of biochar in her region. Education is key to obtaining more funding to make biochar from invasive species in Maryland, to provide benefits for native ecosystems and stormwater management.

Wisconsin Department of Natural Resources

The Wisconsin Department of Natural Resources is using two Ring of Fire kilns to experiment with producing biochar as a forest product. Their first project was to make biochar from hardwood brush cut around the Wilson State Forest Nursery in Boscobel, Wisconsin. The nursery has been using peat to increase the organic matter in the seedling beds on a sandy soil to improve soil structure and water-holding capacity. However, the peat is not only expensive, it is not sustainable, and the organic peat material readily breaks down in the soil and must be replenished regularly. Biochar could be a longer-lasting substitute for the peat. The nursery started a biochar field trial using eight different species of trees and shrubs, to learn more about effective biochar uses.

Figure 12.8. Biochar applied at Wilson State Nursery in Boscobel, WI (*left*). Following application, the biochar was rototilled into the test beds. The plants grew well (*right*). Photos courtesy of Wisconsin DNR

Brian Zweifel, forest products specialist with the Department of Natural Resources, is looking for opportunities across the state to make biochar from sawmill residues and slash from forestry projects, including urban forestry. One special project took place at the Aldo Leopold Foundation. Leopold was the author of the conservation classic *A Sand County Almanac*, a book that has inspired generations of conservationists. A quote from the book is familiar to many: "We abuse land because we regard it as a commodity belonging to us. When we see land as a community to which we belong, we may begin to use it with love and respect."[29]

Zweifel taught the forest managers at the site of Aldo Leopold's cabin in the woods how to make biochar from milling slabs—the rounded sides of a log that are removed to make a square for milling lumber. The forest managers had recently thinned a pine forest planted by Leopold and milled the lumber. Normally, they would have just burned the slabs to get rid of them, but now they can turn them into biochar. The biochar will be used on-site to improve the soil in a fitting tribute to Aldo Leopold and the conservation ethic that he wrote about.

Butte Fire Safe Council and Butte Community College

Paradise, California, site of 2018's devasting Camp Fire, is located in Butte County. Since that terrible event, where 84 people lost their lives and 90 percent of the homes were destroyed, the Butte County Fire Safe Council has been organizing projects to reduce fire risk throughout the county. Stephen Feher, Butte County Fire Safe Council board member and retired professor at Butte Community College, has been advocating to include biochar in the fuels reduction work, with some success.

Prior to the Camp Fire, Professor Feher had established a MESA (Math Science Engineering Achievement) internship project on biochar at Butte College to introduce students to applied research and development and to train them to conduct hands-on, laboratory and field experimentation as a path toward professional development. He won a US Environmental Protection Agency (EPA) grant to fund the program to train students to become part of the biochar industry. The program engaged students in multiple biochar projects to learn how to make, characterize, and use biochar in gardens, greenhouses, and local farms. He brought in biochar experts to inform and interact with the students, and sponsored several biochar demonstration workshops, inviting my company to share our kiln technology with the students and members of the community. Stephen Feher's broad outreach and promotion of biochar to solve problems in California forests and farms has seen great results, including the establishment of biochar contractors in the area. We will look at the work of one of these groups, the Biochar Coalition, in chapter 13.

Biochar Mutual Aid Networks

Following a presentation on biochar to the Oregon Small Woodlands Association, I got a call from Cherie, a small woodland owner who lives a few miles away requesting a kiln demonstration on her land. Cherie called some friends and neighbors to help; I brought a kiln and some tools, and we all pitched in for a few hours to get rid of her burn piles and make biochar for her flower beds. Cherie made a pot of soup and cookies for everyone, so we had lunch while the last load of wood was turning to char. Then we quenched the kiln, raked it out, and loaded the kiln back into my truck and I got home by mid-afternoon. This is how a neighborhood biochar network can bring people together for mutual aid and fun.

Over the past ten years I have lost count of the number of biochar kiln demonstrations I have done at rural woodland properties, small farms, conservation districts, fire districts, and land restoration projects. Now others are picking up the ball and holding their own workshops and training events on how to make biochar in kilns and burn piles. I know the practices are taking off, judging by the number of kilns I have sold (hundreds). In addition, many people are making their own kilns or using the simple conservation burn methods. Those who have rural properties in fire country out west are highly motivated to do the work, and appreciative of a clean-burning alternative to dispose of hazardous fuels that also provides a valuable product for their gardens and landscapes.

According to the National Association of State Foresters, more than half of US forestland is in parcel sizes smaller than twenty-five acres, owned by more than ten million private owners. Judging by what I see in my neighborhood, many of these woodland owners are aging and having trouble managing their land. While some owners sell timber from their lands, or have forestland as part of a working farm or homestead, many woodland owners hold their lands for enjoyment and conservation. Many did not bargain for the consequences of drought and wildfire on the forests they love, and did not realize how much work they would need to do to clear around their homes and thin overcrowded stands that have not seen regenerative fire for more than a century. These people need help. They need help from others to do the work and help to pay for it.

Various agencies are stepping up to provide cost-share funding through programs managed by the federal government, but the amounts are small compared to the need. Everyone wonders how we are going to get this work done, and if carbon credits could help pay for it, seeing as how converting forest slash to biochar is a very solid and easy-to-measure form of carbon removal from the atmosphere.

Illinois Valley Fire Resiliency Oversight Group

I have been involved in forest protection efforts in the Rogue River-Siskiyou National Forest and surrounding areas since 1987. These forests have gone through vast changes in that time and have seen many megafires, including the 2002 Biscuit Fire that raged through a half-million acres of remote, wildland forest. Forest and fire management approaches have gone through many changes as well. One change for the better is an increasingly collaborative approach between community members and the plethora of agencies involved.

A few years ago, after the Forest Service proposed a helicopter-logging timber sale in the Wildland Urban Interface (WUI) in my neighborhood, we convened a group to explore alternatives for light-touch logging that could remove some hazardous fuels and also benefit forest health. After several years of discussion, the project went forward with a sale of poles from the Page Creek Project that made community members happy, although the work did not make a lot of money. The group has since evolved into a large, valley-wide collaborative called the Illinois Valley Fire Resiliency Oversight Group (IV-FROG).

IV-FROG has embraced the priority of hazardous fuels treatments and has committed to a program of public education on clean burning and biochar production. I am looking forward to a chance to bring biochar knowledge to a wider section of our community in the years ahead.

Biochar Camp at Table Rock's Enchanted Forest

It can be challenging to recruit workers for forest restoration and wildland firefighting jobs. Young people entering the workforce are not always aware of the opportunities for good-paying, satisfying work in this sector. For several years now, Brenda Patton's Table Rock Foundation has been hosting an On Ramp Forestry Training Camp at her beautiful Enchanted Forest property to introduce youth to these careers by providing a week of hands-on activities in forestry and ecosystem restoration. Grayback Forestry and Lomakatsi Restoration Project provide training and instruction, and I organize a biochar day where we learn about kilns and conservation burns.

We have also provided biochar programs for the younger group of middle-school-age students. Biochar is a great topic for all kinds of STEM objectives.

Figure 12.9. At this biochar camp we made charcoal from twigs and pine cones in paint can retorts for art projects while engaging in a rocket stove building contest (*left*). Willow sticks processed in the paint can retorts made great drawing charcoal. Art by Trey Hendrix (*right*).

We include the arts as well, with storytelling and creative art projects. We learn about fire, soil, forest ecology, chemistry, and climate. Our activities include making biochar stoves and rocket stoves, making biochar in a campfire, testing the pH of biochar and other substances, looking at soil under a microscope, doing the math to understand carbon and climate, and scavenger hunts in the forest for art materials. Now some of the middle school students are joining our forestry careers camp. The forests are calling them.

CHAPTER THIRTEEN

Counting Carbon

Making biochar has value for soils, water, forests, and climate, but it can be very difficult to monetize such environmental benefits as they often fall outside of the market economy. However, biochar is a robust and measurable form of atmospheric carbon removal. Although carbon markets are still not well developed, the potential exists for biochar production to be at least partially funded by carbon removal credits. The key to monetizing biochar carbon is to find ways to measure and count the impacts of biochar carbon. Combining payments for carbon removal with other market values can make a biochar project viable.

There are many entrepreneurs currently exploring all kinds of opportunities to make biochar from the massive biomass waste streams pouring out of cities, industries, and agriculture. These waste streams include forestry and sawmill residues, sewage sludge, manure from concentrated animal feeding operations (CAFOs), municipal green waste, food-processing waste, and agricultural residues. Biochar, and where possible, energy production, are wonderful uses for material that might otherwise be landfilled or incinerated. A growing industry is diligently working to develop all of these opportunities for biochar production. In addition to developing the technology and securing biomass waste streams for inputs, entrepreneurs also have to develop markets for biochar and its coproducts. To follow new developments in the North American biochar industry, check in with the US Biochar Initiative at biochar-us.org.

Markets for biochar are multitudinous and extremely diverse, ranging from soil amendments, to building materials, to high-tech nano-materials. To learn more about all the amazing uses for biochar and the emerging markets, read *Burn: Using Fire to Cool the Earth* by long-time biochar advocates Albert Bates and Kathleen Draper.[1] They leave no stone unturned in their examination of biochar uses.

Despite the plethora of potential biochar markets, producers badly need mature markets in order to ramp up production. Biochar is very new and most people have not yet heard of it, so there is a lot of education and product development yet to do. In the meantime, carbon credits are already helping to provide a revenue stream to support the developing biochar industry.

How exactly does biochar sequester carbon? Plants convert CO_2 to plant stuff and humans convert plant stuff to biochar that microbes can't easily degrade, removing the carbon from circulation. A tree acts as a suction pump, drawing carbon dioxide down from the atmosphere and storing it, temporarily, in plant tissues. Pyrolysis completes the job, stabilizing the carbon into long-lived biochar. A properly implemented biochar system becomes a one-way pipeline transferring carbon from sky to ground. There is enough knowledge now about biochar properties and longevity to be able to accurately determine the carbon sequestration capacity of a given biochar material, including how much carbon it contains and how likely that carbon is to be oxidized into carbon dioxide when added to soil or used in a product.

In recent years, CO_2 Removal Certificates (CORCs) have become a reality, with several different outfits offering protocols and funding mechanisms to provide payments to biochar producers for biochar they produce based on its content of stable carbon and what that represents in terms of CO_2. A CORC is a tradable digital asset that represents one tonne of CO_2 equivalent that has been removed from the atmosphere. A number of biochar producers are now tapped into this market and are adding CORCs to their bottom line.

But the biochar industry is only half the story because so much biomass is considered "stranded biomass." This material might be in burn piles in a remote forest or on a small farm. It might be the dead branches in a suburban backyard or the piles of grape prunings at a vineyard. It is too expensive, both in terms of dollars and in terms of fuel use, to gather all this up and transport it to an industrial facility for pyrolysis, where it can produce products for market. Without a market incentive, how can we get this material processed into biochar?

Farms, ranches, orchards, and vineyards are able to utilize biochar made on-site as an input to their operations. Whether or not the value of biochar pays for the labor and expense of making it is a question that each operation needs to determine for itself. Often the answer is yes, especially when producers have good information about how to use biochar for the greatest benefit.

Industrial forest landowners have yet to embrace biochar in a big way, but when they do, all of the Biochar in the Woods options discussed in the previous chapters are available to them, along with larger, scaled-up technologies such as full-sized air curtain burners.

Landscaping and grounds maintenance crews for city parks and green spaces have many opportunities to turn green waste into biochar on-site, using biochar kilns or conservation burns. Biochar is a valuable input for landscaping, including tree planting, and improves rain infiltration and soil water-holding capacity. Compacted soils in the urban environment fail to soak up all the water in rain events, causing flooding and erosion. Stormwater management is a huge expense for cities, and there are many current efforts in cities in North America and Europe to add biochar to the urban stormwater management infrastructure.

Biochar as a Verb

No matter how biochar is produced, whether in large-scale industrial pyrolizers and boilers or in small kilns in the woods, it takes a lot of work. That's because biomass is not just another commodity mined from the ground. Biomass is recently living material that is on a path to degradation, is not uniform, has a low bulk density, takes some effort to gather and transport, and needs to be dried, resized, and homogenized before it can be used in pyrolysis equipment.

Biochar takes work—there is no way around it. Biochar as a product with uses and markets is a noun, but biochar as a carbon removal vehicle is a verb. *Biochar (v.): To biochar is to make charcoal from biomass by pyrolysis, gasification, or other thermal methods. The biochar process includes gathering, drying, resizing, and transporting biomass to a biochar production site where it is heated, monitored, and quenched to stop the complete thermal conversion of biomass to ash.*

When considering biochar for a carbon credit, all of these processing steps have to be accounted for, and the carbon footprint of all the steps counts against the carbon removal capacity of the biochar product. The steps also have to be included in cost accounting for the final price of the biochar, whether it is used as a product or a CORC or both. Fortunately, double-counting the value of biochar is OK here. We can get paid for both the value of the product in its end use, and the value of its carbon removal action. This double-counting makes biochar unique in the carbon credit world.

Meanwhile, out in the woods and the fields, we are beginning to train a workforce to do the biochar jobs. Until carbon removal payments are widely available, small businesses and contractors who make biochar in place are being paid for two things: the biochar they produce, and the service of removing the biomass residues. As biochar becomes better known, these values are leading to paying jobs for the biochar workforce.

Arborists and contractors working in agriculture and forestry are starting to include biochar in their services, as an alternative to open pile burning or chipping and hauling. As they develop their methods and complete more jobs, they will learn how to more accurately bid the work and sell the values of biochar to their customers.

Using the example of the Ring of Fire kiln, let's assume a contractor has two kilns and three workers. How much biochar can they make in a day? How much labor will it take? Making biochar takes time. It takes time to load the wood slowly into the kilns. It takes time for the wood to heat up in the kiln, for the wood gas to volatilize and burn, and for the remaining carbon to turn into char. As a rule of thumb, with good, dry fuel, we have found that it takes about 10 minutes to accumulate 1 inch (2.5 cm) of char in a flame-cap kiln. Smaller brush will char faster, and bigger logs and branches will char more slowly. Wet material is very slow, so dry your feedstock!

Using this rule, we can estimate how long it will take to fill two kilns full of biochar. Each Ring of Fire kiln has a total volume of 5.6 cubic yards (4.3 m^3), but we usually don't fill it much past the 24-inch (60 cm) mark. Two feet of biochar in the kiln equates to 3.4 cubic yards (2.6 m^3). To fill the kiln up to 24 inches (60 cm) requires 10 minutes per inch (2.5 cm), or a total of 240 minutes (4 hours). We also need to add time to arrive at the site, set up the kilns, load the initial charge, light it, and get it going. Then there is quenching and unloading, as well as cleanup and loading kilns back into a truck if they are not going to be used again on that site. Total time for the job: 6 hours.

To operate two kilns will typically take three people—one person to load each kiln and a third person to help stage the feedstock close to the kilns. The alternative would usually be to chip the material, renting a chipper at the cost of about $500 a day, plus labor to run it.

How much is the biochar worth? If it is worth $100 a cubic yard, you have produced $680 worth of value in the biochar alone, not counting any carbon removal credits. Does the landowner want it? Or can you take it away and sell it? These are all questions that the contractor needs to answer to decide if the job will pay. There are also questions about burn permits. In some locations it will not be possible to use fire in any form to process woody debris.

Emissions

Many people have wondered if the emissions from making biochar might negate the carbon sequestration value of the biochar. Making biochar on the landscape in flame-cap kilns or conservation burn piles will produce some smoke and other emissions. However, it will be far less than the alternative of burning these fuels in uncontrolled burn piles. Most of the woody debris that we process into biochar would have been burned for disposal anyway, producing visible columns of smoke. Using our eyes only, it is easy to see that our methods produce far less smoke. Particulate emissions are the biggest concern from a human health standpoint, so it may be possible to convince air quality regulators to allow flame-cap kilns for woody debris processing in residential areas. Compared to the smelly diesel emissions and high decibel noise of a wood chipper, residents may find a flame-cap kiln operating in the neighborhood to be far less objectionable.

A couple of years ago, we were operating a Ring of Fire kiln in a neighborhood and we invited our local fire marshal from the Illinois Valley Fire Department (IVFD) to come see what we were doing. When he finally showed up, he said he had trouble finding us because he was looking for a column of smoke and he did not see any. He found us when he heard the sound of the crackling flames. He was happy with the safety of our operation and very pleased to see that we had a hose at the ready for quenching and saving our char. Every year IVFD responds to dozens of burn piles that escape when the wind comes up and blows an ember into dry grass. This often happens because people get tired of tending their slow, smoldering, smoky burn pile and go into the house to do something, taking their eyes off the fire. Our Ring of Fire burns are done in a couple of hours because we always keep a hot flame going to finish the process cleanly and quickly.

Judging by the opacity of our emissions, we are not producing a lot of particulates. However, there are other emissions, including the potent greenhouse gas methane, which can be detected only with special instruments. We now have a few studies that have measured methane emissions from flame-cap kilns, confirming that methane emissions from a flame-cap kiln are very low to undetectable when using dry feedstock.[2] For this reason, when we do carbon accounting, it is very important to measure the moisture content of the feedstock and to only use dry feedstocks that will sustain a hot flame and not smolder and smoke.

Biochar production is optimized when using dry feedstocks and flaming combustion. When feedstocks are wet, the process takes longer and is less efficient, so on purely economic grounds, we would expect that most instances of biochar production in flame-cap kilns will conform to these best practices that result in minimal methane emissions.

If we are sure that methane emissions are minimal, what about carbon dioxide, the greenhouse gas that gets the most attention? We can know from the amount of biochar we produce that we have reduced emissions of carbon dioxide as compared with incineration to ash. It is a simple matter of accounting for the carbon. If we burned the wood completely to ash, all of the carbon would have gone into the atmosphere as carbon dioxide and other gases, but the largest fraction would be carbon dioxide. Clean burning can avoid emissions of the other gases, but carbon dioxide will always be produced by combustion and oxidation of wood. However, if we make biochar efficiently, up to 50 percent of the carbon in the wood is not combusted to ash. It is stabilized in biochar, and does not become carbon dioxide.

Contractor Results

These are exciting times for place-based biochar production. There is a massive need to deal with woody debris across our landscapes, but the question has always been: How can we pay for the work? Several contractors are now up and running with the Ring of Fire kiln, and some of them have shared production numbers, including cost analysis in some cases.

WATERSHED CONSULTING—MARK VANDER MEER

Mark Vander Meer is a restoration ecologist, forester, and soil scientist based in Missoula, Montana. Mark shared results from a 21-acre thinning treatment in western Montana processing ponderosa pine, western larch, Douglas fir, and lodgepole pine. Average tree density in the unit was 1500 stems per acre and most stems were less than 3 inches (7 cm) in diameter. The stand was thinned in 2020 and the material was lopped and scattered and allowed to dry for one year on the ground.

Using four Ring of Fire kilns, Watershed Consulting produced 112.5 cubic yards (86 m^3) of biochar in 85 kiln batches, yielding 5.4 cubic yards (4.1 m^3) of biochar per acre. The labor cost for the burn crew was $22,165 for 403 total hours of work. There were additional costs for project management and administration and an allowance for a water tender and an off-road utility vehicle. Adding up all of the costs, each cubic yard of biochar cost $376 to produce.

The price of biochar is quite variable depending on many factors, including the properties of the biochar, but on the West Coast, we can purchase biochar for about $100 a cubic yard, retail, at the plant gate. Truckload quantities bring the price down quite a bit, but you also have to consider the cost of transporting and applying biochar to a remote forest site. When adding in those costs, the price of $376 per cubic yard, made in place in a remote area and applied during the quenching and raking process, seems quite reasonable.

Some of this cost would have been incurred regardless because the primary job here is the reduction of the dangerous fuels on the landscape. A typical job that includes cutting and building burn piles for disposal pays $1,200 an acre. Only about $400 of that is for the chainsaw work. The remaining $800 an acre pays for gathering up the cut material and carefully constructing the jackpot burn piles. These take skill and work. Sean Hendrix of Grayback Forestry told me that one worker can construct five to eight piles an hour. With twenty to fifty piles an acre, this can be a big effort. After the piles dry for a season, workers come back at the proper time and light them all at once. The cost allowance for this is about $200 an acre. The biggest single cost of the work is building burn piles by hand. Using a kiln avoids this cost, because all the worker has to do is pick up a branch or a small log, carry it to the kiln, and toss it in.

THE BIOCHAR COALITION

One result of the mentoring efforts of Professor Feher in Butte County has been the development of a biochar workforce that is now getting organized as the Biochar Coalition. The first member of the coalition, Ari Mashiah, bought two kilns from me in late 2021. She then spent fifty-six days during winter of 2022 making biochar in the dry, fire-prone western Sierra foothills. Ari mostly worked by herself and started by going around to neighbors who had burn piles and asking them if she could make biochar. After a few days of this, neighbors started to pay her between \$30 and \$60 dollars an hour to dispose of their burn piles by making biochar. I asked Ari what her future goals are:

> My dream is to have people out in our forest clearing brush and making biochar. This is work that our neglected, overgrown, and seriously unmanaged fire threat of a forest deserves. With the time and effort that I have put in to clearing brush and making char with just a couple of friends, we can clear and treat about an acre in three days (about 10 yards of biochar).

Ari has joined up with Tabor Teachout and Ken Scherer to form the Biochar Coalition, which includes a rotating crew of additional helpers, some of whom, like Preston Englant, were part of Steve Feher's Butte College biochar program. Ken Scherer also has an organics company called KO Frass, producing mealworm frass (frass is a polite name for bug poop) and nutrient-rich fermentations for use in soil and for inoculating biochar.

In the winter of 2023, the Biochar Coalition made 300 cubic yards (229 m^3) of biochar in 117 batches from oak, maple, apple, pine, walnut, ash, willow,

Figure 13.1. The Biochar Coalition at work in a walnut orchard. Photo courtesy of the Biochar Coalition

scotch broom, grasses, grapevine, clean lumber, bamboo, and bones. In July 2023, the coalition got a job disposing of walnut trees from an orchard removal in the Sacramento Valley. Orchards and vineyards in California burn thousands of acres of old vines and fruit and nut trees and trimmings every year, mostly in open burn piles under agricultural burning permits. The Biochar Coalition stepped in with four Ring of Fire kilns and made 1,200 pounds (544 kg) of biochar in one day from two rows of walnut trees (about 10 to 12 trees). The workers each made a competitive wage and the landowners were happy to have the biochar. How much more could they have earned if someone had been willing to pay them for the carbon sequestration value of the biochar?

NAPACHAR

Eric Mayer and his partners in Napachar have worked for two seasons now in the vineyards and forests of California's Napa Valley, making biochar from old vines and woody debris. In 2022, as Napachar's first commercial project, they contracted with Silver Oak Cellars to integrate biochar production into their vineyard replacement cycle. They processed the vines into biochar on-site for Silver Oak to return to the soil during vineyard replanting. Using four Ring of Fire kilns, Napachar rented a mini-excavator to feed the kilns mechanically. Instead of using the four kilns as single kilns, they assembled all the panels into two double-size kilns, each with a diameter of more than 11 feet (3.4 m). The larger size helps with machine loading and also with the problem of vines tangled up with wire. When vineyards pull old vines, the wire trellis often gets tangled in the piles and must be burned along with the vines. Having a larger kiln is useful for loading a big mess of vines and wire.

Napachar can produce about 12 cubic yards (9.2 m^3) of biochar per batch in each double kiln. In 2023, they added a Utah Big Box kiln to their tool kit, which added another 8 cubic yards (6.1 m^3) of potential capacity. They appreciate the easier flood-quenching method that is possible with the solid bottom of the Big Box kiln, but they also like the Ring of Fire for jobs where they don't need or want to use heavy machinery.

Eric Mayer has been carefully tracking the costs and inputs for the jobs. In 2022, over the course of eight days, they produced approximately 130 cubic yards (99 m^3) of biochar from 17 acres of pulled vines, distributed around the vineyard in burn piles each approximately 5 yards (4.5 m) in diameter and 2 yards (1.8 m) in height (40 cubic yards [30 m^3] loosely packed). The project consumed 6,400 gallons of local well water (to quench the biochar), 40 gallons of diesel (to power the mini-excavator and water truck), and 10 gallons of propane (to start the kilns). The resulting biochar was incorporated into a 4-acre corner of the vineyard that historically struggled with acidic soil at a rate of 6 tons per acre.

Figure 13.2. Before and after shots of a vineyard where Napachar converted grapevines to biochar for use in vineyard replanting. Photos courtesy of Napachar

In 2023, Napachar worked with Silver Oak, Jackson Family Wines, Tres Sabores, and a private landowner to produce another 170 cubic yards (130 m^3) of biochar over the course of two months, this time from the old vines from 30 acres of vineyard (four separate vineyards), the manzanita and oak branches of 10 acres of fire-scarred forest, and the branches of one ancient oak tree that fell during heavy rains in the winter of 2023. The season consumed another 10,000 gallons of water, 60 gallons of diesel, and 15 gallons of propane. Despite the wet weather (and therefore wetter-than-ideal feedstock), the burns continued to produce meaningful amounts of biochar for use on-site. The vineyard burns, in particular, produced an average of 5 cubic yards (3.8 m^3) of biochar per acre, which the vineyard managers reincorporated into the vineyard soil prior to planting the next generation of vines.

This kind of careful tracking of inputs leads to realistic cost estimating so that the company can produce a profit. The good news is that Napachar has been able to pay its crew of two to three workers a good wage, and they are now bidding jobs at prices that are competitive with chipping.

The other advantage of carefully tracking job inputs and outputs is that this information is necessary for carbon credit eligibility. The vineyards where Napachar works are very interested in knowing their carbon footprint, and biochar is helpful in offsetting the carbon footprint of vineyard operations. This kind of carbon offset is called "in-setting" because it is internal to the company's operations, and the company itself pays for the carbon sequestration. The advantage of in-setting is that there are no transaction costs for verifying and offering a credit on the open market, and no challenges over pricing. The company can then add the carbon they sequestered to their carbon footprint reporting for their customers, who are willing to pay a premium price for the vineyard products, knowing that they are contributing to climate mitigation and soil improvement.

However, it is still necessary, for the integrity of the process, to carefully track and verify the carbon sequestered by the biochar projects. This is where a South African company called Ikhala Impact comes in.

Digital Monitoring, Reporting and Verification with the CHARR App

During our monthly Biochar in the Woods meetings, we have been introduced to Wihan Bekker of Ikhala Impact. His company has developed a carbon accounting protocol with an integrated desktop platform and mobile phone app called the CHARR app that is intended to help a small, low-tech biochar producer gain access to carbon removal credits. The process is known as Digital Monitoring, Reporting and Verification, or D-MRV. The protocol accounts for all of the inputs and outputs of the process. It also needs to make estimates about the amount of stable carbon in the biochar. This is done primarily by monitoring the production process as carbon stability is highly correlated with the temperature of production.

The digital protocol for field data collection and monitoring during the implementation of biochar production is split into three sequential stages:

1. Sourcing stage: Where the initial waste biomass is sourced, collected, and prepared.
2. Production stage: Where waste biomass is carbonized through eligible low-tech pyrolysis for the purpose of biochar production.
3. Application stage: Where the biochar is applied in soils as a long-term carbon sink.

Each stage has a number of key proof points that need to be submitted via the smartphone application in the form of either a photo, video, or questionnaire. The digital software tracks the date, time, and location of the data uploads. The information submitted is then stored in a database for the calculation of the project's net negative emissions. Once biochar carbon removal certificates are issued, they are added to a registry to ensure traceability and transparency.

The key proof points are:

Correct Feedstock Moisture. Wet feedstocks produce more methane, a potent greenhouse gas. Feedstocks need to be dry to be eligible for carbon removal credits.

Flaming Combustion Throughout the Burn. If the feedstocks are dry, properly sized, and loaded correctly, flaming combustion will predominate during the burn. Flaming combustion ensures that emissions are minimal, as most pollutants and smoke burn up in the flame. Flaming combustion also indicates that the temperature of formation of the biochar is high. High temperature is correlated to stable carbon that will not degrade for hundreds of years.

Mass of Biochar Produced. We need the dry mass of the biochar we produced in each batch so we can attach a carbon removal number to it.

Number three is the tough one. It took some time to figure out how to measure the actual amount of biochar that we produce in one batch in the Ring of Fire kiln. It is easy enough to measure the volume of biochar we make, but it is not so easy to weigh several cubic yards of biochar out in the field. Worse than that, once you quench it with water, you then have to figure out how to subtract the weight of the water. However, if you know the volume of the batch, you only need to know the average dry bulk density of the biochar to calculate the mass of the batch.

Measuring Dry Bulk Density in the Field

Bulk density of biochar is a tricky thing to know because it is always changing, dependent on particle size and moisture content. Thanks to Tabor Teachout of the Biochar Coalition, we found a usable method for doing field measurements. At the end of the burn, before we add any water to the kiln, we level the hot char with a rake and measure the distance from the top edge of the kiln to the top of the biochar in the kiln. That is how we calculate the total volume of biochar.

Then we shovel hot coals out of the kiln into a metal bucket of known volume and quickly weigh it using a simple hanging scale. We can repeat this a couple of times to get an average value. This gives us our bulk density.

Here is a summary:

Measure the Volume of Biochar in the Kiln

1. Before quenching, use a steel rake to level the hot, glowing coals in the kiln.
2. Place a measuring stick vertically in the kiln, against the kiln wall, so that one end touches the level char. Take a picture of the measuring stick that shows the depth of the char in the kiln. Repeat this measurement in two more places around the kiln and record it with a picture.

Measure the Biochar Dry Bulk Density

1. Weigh a metal bucket to get the tare weight using a hanging scale. Take a picture to record the weight.
2. Immediately after making the volume measurements, and before quenching, fill the metal bucket with hot glowing coals shoveled from the kiln.
3. Weigh the bucket of hot coals and take a picture to record the weight. Repeat the sampling procedure two more times, taking samples from different parts of the kiln and recording the value with a photo.

Here are the calculations you need to do to find out the weight of the biochar made in a particular batch:

After subtracting the weight of the bucket from all three weight measurements, average the three weights to get one average weight. Then divide the average weight by the volume of the bucket. For instance, let's say we used a 2-gallon bucket, and the biochar average weight was 2.2 pounds, then 2.2 pounds / 2 gallons = 1.1 pounds per gallon.

Let's convert that to cubic feet. One gallon is 0.13 cubic feet. Divide 1.1 pounds by 0.13 to get a dry bulk density of 8.5 pounds per cubic foot. There

Figure 13.3. Compute the total volume of a biochar batch by measuring the height of the char in the cylindrical kiln (*left*). Determine the bulk density of a batch by weighing a sample of hot char in a bucket of known volume (*right*).

are 27 cubic feet in one cubic yard, so a dry cubic yard of our biochar will weigh 8.5 × 27 = 230 pounds. The bulk density of the biochar is 230 pounds per cubic yard.

Meanwhile, we found that the volume of the biochar in the kiln was 3.2 cubic yards. We multiply the volume by the bulk density to find that we made 3.2 × 230 = 736 pounds of biochar. The CHARR app will tell us how much actual carbon removal we have accomplished with our 736 pounds of biochar by converting the carbon we stabilized to a CO_2 equivalent and subtracting the carbon footprint of our operations.

A quick and dirty estimate we could use would be to multiply the 736 pounds by a factor of 2 to get 1,472 pounds of CO_2 equivalent, or two-thirds of a metric tonne. If every person in the United States made about twenty batches of biochar a year, we could offset almost all of our carbon emissions. Let's get to work!

Using the CHARR App and Funding Biochar Production Activities

Out in the field, making biochar, you need to approach the process methodically and organize your work so you can easily make the measurements and record the pictures and video needed to confirm that you did the work and produced the product. We have published an open-source paper in the *Journal of Visualized Experiments* that takes you through the process step by step, providing both written and video instructions.[3] The process verification points for a typical batch of biochar made in the Ring of Fire kiln are listed here:

1. Report feedstock type.
2. Take a picture of feedstock pile to show feedstock size.
3. Take one a picture showing moisture meter reading from largest piece of each feedstock species to show feedstock moisture.
4. To show ignition, make one 30-second video showing start of kiln burn, recording time of start. Video should show a strong flame cap has developed.
5. Record three 30-second videos showing strong flame presence during the first three hours of the burn to verify production temperature.
6. To show kiln volume, include three pictures of a measuring stick in the kiln to show the level of the finished char in the kiln at three locations. Measured distances from the top of kiln to the char are averaged to one value for calculations.
7. To record bulk density, include one picture of the hanging scale showing the empty weight of the bucket. Include three pictures of the scale showing the weight of the char and bucket taken from three locations in kiln. The three weight measurements are averaged to one value for calculations.

Once all of these verification points are entered into the phone app, the data is uploaded to the main CHARR platform where it is processed and the protocol calculates a carbon removal value for each batch of biochar. The program can produce a report that is shared with the landowners to use however they like.

The D-MRV results provided by the CHARR platform can be instrumental in securing finance or funding for biochar production projects. These results, which quantify the carbon removal benefits of biochar production, serve as a reliable and verifiable data source that can be presented to potential funders. In the evolving landscape of carbon removal, such data is invaluable. It provides a solid foundation for projects to demonstrate their environmental impact and sustainability, thereby attracting funding from companies, governments,

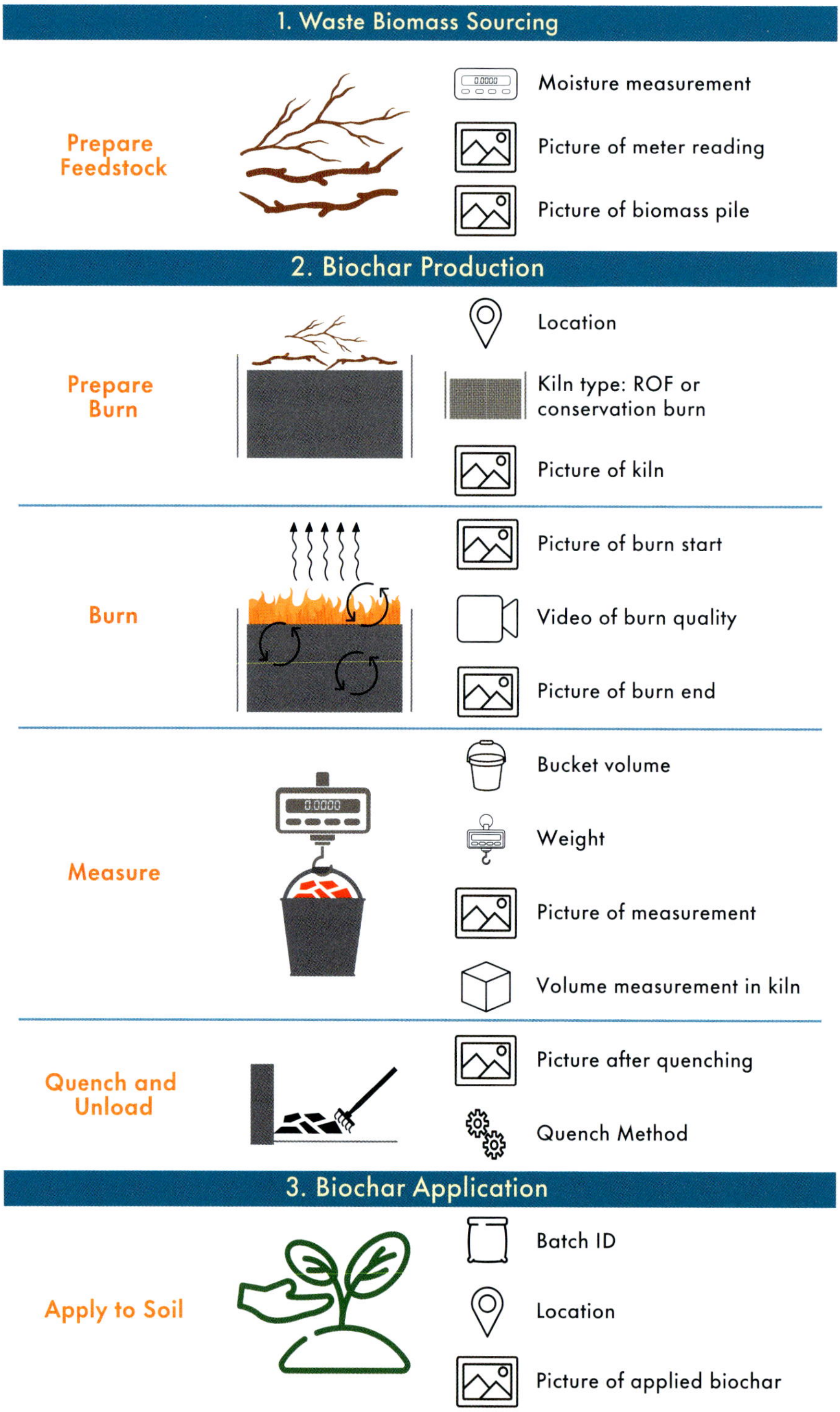

Figure 13.4. Flowchart showing the steps for Digital Monitoring, Reporting and Verification using the CHARR app.

municipalities, conservation groups, and other entities interested in supporting environmentally sustainable practices.

However, navigating this changing landscape to find funding requires a strategic approach and selling carbon credits is one of the most complicated models available. An alternative approach is to engage with local entities that can fund biochar production at a project level, where they have a mandate to reduce wildfire risks, reduce greenhouse gas emissions, or improve soil health. This carbon "in-setting" model allows these entities to directly contribute to the biochar carbon removal in their ecosystem or value chain, while also reducing the costs associated with due diligence.

The ultimate goal for biochar projects should be to collect funds for both the product and the services of biochar. We also need to make sure that funders and investors understand the full range of services that "biochar as a verb" supplies. Not only does the biochar process convert waste into value and sequester carbon, the biochar product provides additional services once it is in the soil. Purchasers of biochar products know this because that is why they want the product. They know it will hold water and nutrients in soil and benefit soil life and health in many ways. What they might not be considering is the long-term effect biochar has on the soil microbiome and plant growth. It is hard to quantify, but biochar strengthens the belowground soil ecosystem, improving its ability to sequester carbon in soil.

Biochar primes the soil carbon pump by improving the productivity of the site so more biomass grows, increasing the pool of soil carbon in the living biomass, and creating a virtuous cycle of carbon drawdown. Furthermore, biochar in soil reduces soil emissions of greenhouse gases. Researchers at Zhejiang University in Hangzhou, China, examined reports of biochar impact on soil emissions from many different soil types. They concluded that biochar could be precisely targeted to soil applications that would be most likely to respond with reduced greenhouse gas emissions, increasing the carbon sequestration potential of biochar by an average of 43 percent.[4]

Carbon Wanted: Dead or Alive

Carbon accounting frameworks often focus on stocks of carbon—the carbon pools in the atmosphere or the soil that are more or less stable. But let's face it—stable carbon is dead carbon. Living carbon is dynamic and mobile. Living carbon is part of the flow of carbon as it cycles between earth and sky. While we want to reduce the pool of carbon in the atmosphere, the solution is not to merely increase the pool of inert carbon in the soil. Biochar may not be easily degradable, but it is active in the sense of a catalyst. It is a catalyst that promotes life, increasing the dynamic carbon pool of the living biosphere that continues to draw down more carbon as plants, animals, and microbes grow and flourish. Carbon bean counters have a hard time with this because it's hard for them to count on living things staying alive. Increasingly, planting trees for carbon credits is frowned upon because it is hard to ensure that the trees won't be cut down or burned up. But when you think about it, isn't that what we are here for? To promote living things?

I sometimes wonder what would have happened if humans had never evolved and developed to become the great liberators of fossil carbon that we are. Fossil carbon is dead carbon from ancient life-forms that has been sequestered from the living flow of carbon cycles for millions of years. The fact that we have released this stock of carbon back into the carbon cycle may benefit life in the long term by increasing the amount of carbon that is available for living things to use. Who knows? It is an interesting thought.

While we should do our best to not add more CO_2 to the atmosphere, I think it is more productive to consider the excess CO_2 that is already there as a resource for us to use rather than a pollution problem we need to clean up. If we look at it as a pollution problem, our governments will choose solutions like funding dubious megaprojects to pump CO_2 gas into underground storage reservoirs, benefitting no one but big corporations.

CHAPTER FOURTEEN

Return to the Garden of Eatin'

The ultimate goal of farming is not the growing of crops, but the cultivation and perfection of human beings.

—Masanobu Fukuoka

When I learned about terra preta back in 2007, my first thought was: "Aha, permanent compost!" As a gardener aspiring to produce all of my own vegetables, I had done my share of barn cleaning, manure hauling, and compost turning, and the idea that I could reduce the amount of backbreaking work I had to do was very appealing. Replacing soil carbon every year in even a small garden like mine is a huge task.

Figure 14.1. The Garden of Eatin' is what we call our vegetable garden that feeds two families with plenty of extra to share and trade with neighbors.

As I have learned over the past couple of decades, this biochar promise holds true. My garden is better than ever, truly a Garden of Eatin'. My garden partners and I are able to produce almost all of the vegetables that our families eat, plus lots of extra to give away and trade. Growing our own food is important to us for many reasons. First, it is hard to get good-quality produce in our area and it is expensive. Grocery store produce may have pesticide residues and it is grown on depleted soil so it lacks nutrients.

Those are old problems. Now we have to worry about new ones like genetic engineering and even the possibility that produce can be turned into vectors for pharmaceuticals. A report from the University of California describes a $500,000 project at the university to investigate the possibility: "The future of vaccines may look more like eating a salad than getting a shot in the arm." UC Riverside scientists are studying whether they can turn edible plants like lettuce into mRNA factories.[5] I thought perhaps this was just one misguided study, but I found another news report describing parallel research at the University of Ottawa.[6] I want to control what I put in my body. I want to eat the food that Nature evolved me to eat, not food that is genetically engineered by humans who have no ability to know the consequences of their manipulations.

Big food and big pharma have a lot of bad ideas they would like to sell us because it makes a profit for them. Cheap, nutritionally hollow food, produced with chemical fertilizers and toxic pesticides, is mining soil carbon the way we mine fossil carbon for fuel. The loss of carbon leads to deterioration of soil structure with further impacts on erosion, compaction, water-holding capacity, and soil biota. In dry areas, these impacts may compound and result in desertification. Mechanized agriculture rapidly compacts soils that are carbon-depleted. In the wealthier regions of the world, the loss of carbon is

Figure 14.2. It's dinner time. Do you know what's in your salad?

being masked by increasing applications of NPK fertilizers and other inputs, but this leads to further problems such as soil acidification, loss of soil biota (with impacts on plant health and disease), water pollution, and increased dependence on chemical fertilizers and pesticides. In the poorer regions, desertification is rapidly increasing.

The Earth can provide abundantly for our need but not for our greed. Greed and profit have led to unprecedented soil loss and land degradation, as well as to the ongoing problem of global warming. As a result, food security is in danger everywhere.

Scope of Land Degradation

Land degradation is a huge global problem affecting all continents.[7] According to the UN Food and Agriculture Organization (FAO), hotspots for land degradation are along the west coast of the Americas, across southern Europe and North Africa, the Sahel and the Horn of Africa, and throughout Asia. More recently, out of 1600 million hectares of cultivated land, the FAO finds that 34 percent of land is in a critically degraded condition and is being further degraded. Only 11 percent of the global land surface is considered to be suitable for producing crops. This small slice of land must feed the current and future population. The amount of arable land cannot be increased without causing further degradation, so the only hope of maintaining production is to rehabilitate the degraded lands.

Land degradation is defined as the loss of productivity or utility. In the context of productivity, land degradation can result from a mismatch between land quality and land use.[8] Some lands are more resilient than others to the impacts of farming, logging, grazing, and other uses. Land degradation includes physical, chemical, and biological processes. It begins when natural land is cleared for agriculture or when domestic grazing animals are introduced. The first impact is a loss of carbon both aboveground and belowground in the form of plant biomass, soil biota biomass, and decomposing soil organic matter. Globally, the loss of soil carbon is huge. Most agricultural soils have lost between 25 percent and 75 percent of their soil carbon in the last five decades.[9]

Methods of water use are related to the degradation of soils and watersheds. Irrigation can cause soil salinity by depositing salts in the root zone of plants that are difficult to flush out. Conversely, agricultural chemicals are liable to leach out of the root zone into groundwater, causing eutrophication and toxic contamination of water bodies.

Industrial processes and development also degrade land. Air and water pollution, mining, and urbanization have all contributed to degrading the land that could otherwise be used for agriculture or to provide ecosystem services.

Fixing Soil

One of the first and most important tasks for biochar is to help repair the vast amount of degraded and damaged land. It is natural to think of biochar as an agent for restoring degraded land because of its history of use. Native people in the Amazon rainforest mixed charcoal with human waste and kitchen waste to form the highly fertile soils known as terra preta. These soils are a prime example of the relationship between marginal land, agriculture, ecosystem services, and degraded land.

Tropical rainforests are regarded as land that is marginal for agriculture. Soils are highly leached and low in soil organic matter, yet they support biologically diverse forest ecosystems that provide important services such as water, carbon storage, and climate regulation, among others. Tropical lands are marginal for agriculture but perfectly suitable for biologically diverse forests. Converting such lands to agriculture on a large scale inevitably leads to a long-term loss of carbon storage and ecosystem services.

On a small scale, however, native Amazonians were able to use these lands for agriculture without degrading the soils because of their biochar amendments. This example should be used not to justify clearing more primary forest for agriculture (these forests are badly needed for providing ecosystem services) but as a model for restoring the millions of hectares of soils that have already

Table 14.1. The Impact of Biochar Applications on Soil Degradation Problems

Soil Degradation Problem	Biochar Beneficial Impact
Loss of nutrients	Adsorbs N, increases CEC, reduces leaching.
Soil compaction	Loosens soil by reducing bulk density and soil tensile strength. Promotes soil aggregates.
Soil erosion	Promotes root growth and mycorrhizal fungi, holding soil in place.
Droughty soil	Increases water-holding capacity.
Loss of soil biota	Supports soil biota by providing air, water, nutrients, and refuge from predators in biochar pore spaces.
pH imbalance	Buffers pH over a wide range.
Salinization	Forms moisture-retentive upper soil horizon in root zone. Loosens soil and encourages salt leaching.
Toxic contamination	Immobilizes and prevents plant uptake of heavy metals and persistent organic pollutants. Enhances degradation of some organic pollutants.

been degraded by human activities. Biochar can help restore degraded land and improve marginal land for agriculture and ecosystem services. Table 14.1 lists some of the areas where biochar can improve soils and summarizes the mechanisms by which it acts.

In addition to improving soils in general by providing carbon, water-holding capacity, and habitat for soil life, biochar can be helpful with some specific soil problems that we face if we are to fix our most damaged soils.

Soil Compaction and Erosion

Biochar reduces the bulk density and tensile strength of soil while encouraging root growth. It helps soil to form aggregates of soil organic and mineral particles that result in improved soil structure for root penetration and water-holding capacity. These properties can help alleviate soil compaction and present a defense against erosion as root growth takes hold. A study of highly erodible soils on the Loess Plateau, China, found that biochar enhanced the formation of soil aggregates of silt loam soils, potentially increasing soil stability and reducing soil erosion.[10]

pH IMBALANCE—ACID SOILS

On acid soils, biochar can have a liming effect while also increasing cation exchange capacity.[11] A study of phosphorus uptake on acid soils found that the liming effect of biochar resulted in greater plant uptake of phosphorus.[12] USDA researchers working with sandy acid soils in the US Coastal Plain region found that biochar increased soil pH, soil organic carbon, calcium, potassium, manganese, and phosphorus.[13]

pH IMBALANCE—ALKALINE SOILS

Even when the biochar itself is alkaline, it can buffer alkaline soils because of acid functional groups on biochar surfaces created by soil microbes.[14] An incubation study using soils from the Loess Plateau in China found that adding an alkaline biochar to alkaline soils reduced pH over several months. Some moderation of the soil salinization process was also observed.[15]

SALINIZATION

Biochar can alleviate soil salinization through several mechanisms. By improving soil water-holding capacity, it can form a moisture-retaining upper soil horizon in the plant root zone.[16] A two-year field study of biochar in salt-stressed cropland in China found that biochar significantly improved seed germination, crop yield, and soil quality.[17] Biochar alleviated soil compaction, encouraging the leaching of salt. It may also have immobilized some salt, making it unavailable to plants. The beneficial effects of biochar persisted and even increased in the second year of cropping, without any additional biochar application. A meta-analysis confirmed that these results were replicated by other studies, and furthermore that biochar increased the amount of microbial activity in salt-affected soils resulting in improved plant growth.[18]

TOXIC CONTAMINATION

A multitude of studies have shown that biochar can sorb many different kinds of contaminants and either degrade them or immobilize them.[19] In some cases, the liming effects of biochar can increase mobility of arsenic, but one study showed that even though arsenic mobility increased, its uptake to plant tissues was reduced.[20] Biochar can revitalize soils contaminated with lead.[21] This has important implications for the use of soils for urban agriculture where lead contamination is a major limitation.[22] Other types of soil contamination can also be ameliorated with biochar, including PCBs,[23] heavy metals,[24] and pesticides.[25] Biochar can also promote the degradation of pollutants. A study that looked at biochar as a tool for cleaning oil spills found that biochar reduced volatile emissions and accelerated the microbial degradation of volatile hydrocarbons in a sandy soil.[26] Mine tailings are a long-term source of pollutants that need to be remediated through revegetation. Biochar has been shown to be effective in promoting revegetation and reducing the leaching of toxicants from mine tailings.[27]

Dominion or Reverence?

This is not our first rodeo with ecocatastrophe. We humans have been here before. In his book *Out of the Earth*, soil scientist Daniel Hillel gives us a history of human agriculture and soil degradation.[28] This history is written on the land and easy to see, for a soil scientist. Hillel says, "The poor condition of the 'Fertile Crescent' today is due not simply to changing climate or the devastations caused by repeated wars. . . . It is due in large part to the prolonged exploitation of this fragile environment by generations of forest cutters and burners, grazers, cultivators, and irrigators, all diligent and well intentioned but destructive nonetheless."

The Mediterranean is a "haunting example of soil abuse on a large scale," he says, and this may have been the reason why "the Phoenicians, Greeks, Carthaginians, and Romans, each in turn, were compelled to venture away from their own country and to establish far-flung colonies in pursuit of new productive land."

There were, he says, a few societies that figured out the keys to soil sustainability, naming China and parts of Southeast Asia, as well as the "wetlands-based societies of Meso-America and South America." We can add the Amazonian societies who created terra preta to this list.

Hillel wonders if the two contrasting accounts of Creation in the Hebrew Bible might portray different approaches to agriculture. In the first creation story, God makes man "in his own image" and instructs him to "be fruitful and multiply, and fill the earth and conquer it." In this version of the story, every other form of life is created for the benefit of man. This is the concept of dominion.

In the second chapter of Genesis, the story is retold, but this time, God makes man—Adam—out of earth (*adamah*) and instructs him to live in the Garden of Eden and to tend and protect it as God's steward. The next thing that happens is the temptation to eat of the tree of knowledge of good and evil and become like gods. Adam and Eve do the deed and get expelled from Eden to wander the Earth in sin and travail.

These stories and history should serve to remind us that, while we may seek self-reliance and independence from exploitive governments and economic systems, we will always be dependent on the Earth itself for our life and sustenance. To be truly free, we will need to become not as gods but as humble stewards of God's Green Earth. This was the Jeffersonian vision of the "yeoman farmer," a class of independent farmers who would settle the New World and have a new chance for freedom and independence away from the exploitative aristocrats of Europe.

Of course, what really happened was that the settlers were colonizers who decimated the native people while eventually being reduced to serfs again themselves. The loss of family farms and consolidation under corporate agriculture in the United States over the past fifty years has been devastating to both people and the land.

Farming almost always degrades soil, and modern farming practices are the most destructive of all. There are two schools of thought about how to fix this, paralleling the two Genesis stories.

One school wants to industrialize agriculture even more and bring it under centralized control in the name of sustainability and climate change. They would eliminate meat and milk, for instance, and grow protein in factories. But centralized and highly mechanized industrial agriculture, and the global supply chains it depends on, have no future. The reason is simple: energy.

The world will likely reach peak oil production—from all forms of oil: conventional (light, sweet crude) and unconventional (tar sands, oil shale, heavy oil)—sometime before 2030. There will still be oil left in the ground, but the cost of extracting it will become increasingly unaffordable as ERoEI (energy return on energy invested) plummets. It is all downhill from here, in terms of the energy that will be available to us to power our civilization going forward. Wind, solar, and other non-fossil-fuel energy sources cannot fill the gap.

Actuary Gail Tverberg explains all the reasons why in an article titled "Models Hide the Shortcomings of Wind and Solar."[29]

Economist Nate Hagens is a current leader in the discussions around our energy future. He predicts a "great simplification" as energy costs go up and availability goes down. He has an illustrative example in agriculture: how to milk cows. Manual milking uses 120 hours of human labor per cow per year. Semiautomated machine milking uses 27 hours of human labor per cow per year, and fully automated milking machines require 12 hours of human labor per cow per year. The first method uses no electrical energy, the second method uses 1.1 megawatt-hours per cow per year, and the fully automated method uses 3 megawatt-hours per cow per year. A fully automated dairy with 5,000 cows would use 15,000 megawatt-hours per year just for milking machines. The old way of milking cows used only human energy, and the return on the human energy invested was always positive. With cheap electricity available, automated milking improved the economic return on investment, but as energy costs go up, the returns diminish and even become negative.[30] The conclusions are inevitable. There will come a time in the future when only human energy will be affordable for many tasks.

The other school of thought on how to fix agriculture wants to relocalize farming and supply chains, to move slowly and deliberately down the off-ramp of energy descent while building properly scaled agriculture and industry to support people the best we can. Biochar has an important role to play in this approach to bioregional, local, and sustainable agriculture, energy, and waste management.

Every human civilization on record has collapsed when it has exhausted its resource base. Unlike hunter-gatherer cultures that persisted for 100,000 years, agriculture has only been around for about 10,000 years and we still have not gotten it right.

Alternatives to Dominion

Why has every agricultural civilization either gone bust or been conquered by another civilization that was more aggressive? It's tempting to call it a spiritual problem—we are guilty of hubris, of trying to be gods. Maybe that's it, or maybe it's just because we are like every other animal that will exploit every resource it can reach to feed itself and grow and multiply. Whatever the case, as we come up against the hard and fast limits to growth, there will be pain, followed by adaptation to our new, low-energy environment.

Is there any chance that we could get it right next time and give up the idea of dominion as we evolve a new social order beyond the Petroleum Age? I learned about biochar from others in the peak oil community of analysts and writers that emerged around 2005, when the world reached a peak in conventional oil production and people were talking about peak oil for the first time since the 1970s. One example that many of these analysts and writers used was the Edo period in Japan.

The article "Japan's Sustainable Society in the Edo Period (1603–1867)" describes a time when Japanese society closed itself off from the rest of the world and refused to accept any of the new technologies from Europe, like firearms and coal-powered machinery. Reportedly, it was a time of peace, prosperity, and cultural flowering, with a stable population. This was no primitive society: "The population of Edo, at the time the largest city in the world, has been estimated at 1 million to 1.25 million people. In comparison, London had about 860,000 people (1801) and Paris about 670,000 (1802)."[31]

I have always been fascinated by the adaptation of isolated island nations like Japan. In the larger world, over the past thousands of years, warlike cultures could seize goods from their neighbors, while exploited populations could migrate elsewhere to escape the predation, if only temporarily. A remote island culture must learn to live within its limits no matter what. It must learn the lessons of sustainability.

As we reach the absolute limits of the Earth, we can now see that Earth is an island in space. The limits are clear, and have been for a long time, but ever since I was a child in the 1960s, our culture has only been able to spin endless tales of outlandish science fiction and utter fantasy about the future, with very few willing to mentally cope with the physical reality of limits.

One of the books that most influenced my thinking as a young person is *Muddling Toward Frugality* by Warren Johnson, published in 1978.[32] What I love about this book is Johnson's upbeat attitude toward a future with physical limits. On the one hand he states: "Increasingly, the situation regarding resources will determine the shape of the future more than our political, economic, or social aspirations." But on the other hand, he thinks this may be a blessing in disguise. The difference between the past and today is the difference between traditional societies constrained by religion and ethics, and societies based on exploitive capitalism and endless economic growth. He wonders:

> Whether one system offers more freedom than the other is a knotty question. Philosophers still argue about what freedom really is, but it seems entirely possible that the opportunity to be oneself within the constraints of an unchanging but supportive traditional culture may be a truer freedom than to live in the fluid and often threatening situations we find ourselves in today.

One of the most reassuring messages in the book is that there is no point in trying to persuade everyone who is caught up in the project of dominion. The call of materialism is too seductive and not one of us is immune to the desire for comfort and abundance. It makes more sense to spend our time diligently preparing for a more resource-constrained future. When those limits bear down upon us, as they are doing right now, humans will adapt. We will muddle through and evolve local solutions created by communities for themselves. Warren describes how this will go:

> Later, when the reality of the new trends is understood and reluctantly accepted, people can be expected to start to reorganize their lives. . . . No one likes to have his job and the well-being of his family dependent on vast, impersonal market forces over which he has no control. People will begin to experiment with various forms of barter, cooperative arrangements among friends, and larger scale cooperative organizations—all designed to give people more control over their own lives. In the long run, as the momentum of modern economic society weakens, it can be expected that the market system will decline from the dominant position it has held since the eighteenth century and slowly be relegated to the subordinate place it has held in most societies—of facilitating the exchange of goods at the Saturday market, but not of ordering all of society.

Masanobu Fukuoka, author of the beloved book *One-Straw Revolution*, mourns the passing of the "one-acre farmers" of Japan, who only a few decades ago made up 70 to 80 percent of the Japanese people. He said:

> Farmers at the time were said to grow a hundred crops. In the paddy field and vegetable gardens, they raised rice, barley, and other grains, as well as sweet potato and many different vegetables. Fruits ripened on trees next to the farmhouse, which was surrounded by a shelterbelt. A cow was kept under the same roof and chickens ran about loose in the farmyard, protected by a dog. A beehive hung from the eaves.
>
> All peasants were totally self-supporting and enjoyed the richest and safest possible diet. That they are seen as poor and hungry may well reflect the envy of modern man more than anything else. People today have never had the experience of living independently by their own devices, so they [understand] neither spiritual nor material poverty and abundance.[33]

Hold this vision in mind and do not despair for the future. We can return to the Garden of Eatin' and we can start with what is right under our feet, wherever we are. The good news is that the carbon we need in soils has not gone anywhere—it is still here in Earth's atmosphere. Every one of us has the ability to work with plants to help them draw down this carbon to grow beautiful gardens again.

When it comes to climate change, let's change our perspective from "too much carbon in the air" to "not enough carbon in the soil." We are good at being miners and exploiting resources, so let's mine the air and stash the carbon back in our soil.

I hope this book has shown you some ways that biochar, and the example of the ancient charcoal cultures found in Brazil, Japan, and many other parts of the world, can help you in your quest for self-reliant sustainability, better food, and better health. And more than that, I hope you will find that biochar is a useful tool to help fulfill our mission as stewards of the Earth.

Acknowledgments

The biochar community of researchers, inventors, entrepreneurs, do-it-yourselfers, enthusiasts, and promoters is a fine example of humanity at its best. I wish to thank the hundreds of people who have answered my questions at forums, at conferences, and by email communications, generously sharing their knowledge and experiences. I am especially grateful to all the Carboneers who have spent countless hours with me out in the forest, tossing wood into flaming kilns while we plot the Biochar Revolution. It has been a joy to know you and to work with you for a better future for us all.

An extra-special thanks to George and Ruth Ann for your support and encouragement.

APPENDIX A

Plant Bioassays to Evaluate Biochar Compost

The plant bioassay to test composts and soil mixes is a two-week germination and growth test using cucumber seeds. The basic procedure is to mix the target material with a plain, peat-based potting soil at the desired rate and plant the same number of seeds in each pot. Trays of pots are placed in a controlled environment, like a greenhouse or a cannabis grow tent, and grown for two weeks. A single experiment will compare several different treatments consisting either of different amendments to a standard soil or pre-mixed soil blends. We always add a control using plain potting soil.

At the end of the growth period, we count the number of germinations (low germination rate is a strong indicator of phytotoxicity) and the number of secondary leaves. The number of secondary leaves may indicate different levels of growth-promoting hormones, nutrients, or other constituents in the target amendment. Then we cut, dry, and weigh the biomass of each treatment. The biomass weight will tell us something about nutrients available for growth.

The plant bioassay is useful for comparing the effects of different amendments or soil blends on plant growth. A complete protocol is outlined below.

Materials

Plain peat-based potting soil with no added nutrients, such as Sunshine Mix #4 with perlite: used for the control and the base soil if testing amendment treatments

A number of 4-inch (10 cm) round plastic pots

A package of 16-ounce (475 ml) clear plastic drink cups to use as humidity domes

Greenhouse trays

Labels, tape, and indelible markers
1,000-milliliter beaker
100-milliliter beaker
Clean bins for mixing soil (we use bus trays from a restaurant supply store)
Mesh bags (we used a nylon-net bath scrubber—unfurl it and it is one long mesh tube that can be cut into sections to use as bags to contain the plants for drying)
Food dehydrator
Scale accurate to 0.01 grams
Controlled growth environment such as a greenhouse or growth chamber

Preparing the Treatments

1. Determine your treatments (what you will test), plus a control using plain potting soil, and label all pots with a treatment code and replication number. Use at least three replications. Five is better.
2. Measure out the soil.
   a. Fill one pot with plain dry soil and tap the pot gently against the table surface to pack lightly, filling to within 1 to 2 centimeters from the top.
   b. Pour soil into 1,000-milliliter beaker and record volume.
   c. Then weigh that volume = *a*.
3. Fill all the *control* pots with soil.
   a. Multiply weight *a* times number of replicates (usually 3 to 5).
   b. Weigh that amount of soil and place in bin—this will be soil for control pots.
   c. Measure a volume of water equal to half the volume of the soil and mix with soil to moisten. Let stand for 10 minutes, then fill pots evenly.
4. Fill treatment pots for soil blends. If you are comparing already mixed soil blends, simply measure, weigh, and moisten each soil blend as in step 3 and skip to step 6.
5. Fill treatment pots for testing amendments.
   a. Determine the application rate for each amendment you want to test. In this example we use 20 percent by volume.
   b. Measure and weigh enough plain soil for each treatment as in step 3, but do not moisten it yet. Place the soil for each treatment in a separate bin.
   c. Multiply the volume of one pot by 20 percent for the amendment volume. For example, if the pot volume is 300 milliliters, then 20 percent of that volume is 60 milliliters.
   d. Multiply the amendment volume by number of replicates.

   e. Measure the total volume of amendment material into a beaker. Tap to pack lightly. Place the total volume of amendment material in the bin with the dry soil.
   f. Mix soil and amendment thoroughly.
   g. Measure a volume of water equal to half the volume of the pots and mix with soil to moisten. Let stand for ten minutes, then fill pots.
   h. Repeat for the next treatment.
6. Plant seeds.
   a. Place 11 cucumber seeds from the same seed packet evenly spaced on the surface of each pot. Keep seeds away from sides of pot.
   b. Place about 50 milliliters of moist, un-amended soil on top of the seeds and press to spread the moist soil evenly across the top of the pot, covering all the seeds with at least 5 millimeters of soil.
   c. Add 60 milliliters of additional water to each pot to make sure that soil is completely saturated.

Growing the Plants

7. Prepare plants for growth chamber or greenhouse.
   a. Place a humidity dome (plastic drinking cup) on each pot and secure with tape if needed.
   b. Place pots on trays, randomizing the placement of pots within each tray or block.
8. Growth phase.
   a. Determine lighting schedule and set up timers if using artificial light.
   b. If growing in greenhouse, supply heat mats if needed.
   c. If growing in growth chamber, adjust exhaust fan temperature control for normal room temperature.
   d. Grow plants for 14 days. Do not open humidity domes or add any more water.

Figure A1.1. Pots from all the treatments are randomly placed on trays in the growth chamber (an inexpensive cannabis grow tent).

Figure A1.2. After the 14-day growth trial, sort the pots by treatment type. Take photographs to record leaf color and growth habit.

Figure A1.3. Count the number of germinations and the number of well-formed secondary leaves in each pot.

Figure A1.4. Snip off the plants at the soil level, then place all plants from a single treatment in one mesh bag and dehydrate. You will have one mesh bag for each treatment.

Figure A1.5. Weigh the total dried biomass from all of the pots for each treatment type.

Evaluating and Reporting the Results

9. Evaluate results.
   a. Sort the pots by treatment type.
   b. Count number of germinations in each pot and record.
   c. Count number of well-formed secondary leaves in each pot and record.
   d. Take photographs of trays to record information such as leaf color.
   e. Snip off aboveground vegetation at soil level for all of the pots in each treatment (including control) and place in one labeled mesh bag for each treatment.
   f. Place mesh bags of vegetation in drying oven or food dehydrator and dry for 24 hours at 105°F (41°C).

g. Remove dried vegetation from each bag and weigh biomass for that treatment.
h. Record total biomass for each treatment.

10. Report your results. Make a table to record your data, take pictures, and describe your methodology. The previous photos and data from the following example were provided by Grant Scheve, vice president of business development for Agra Marketing Group (Ripon, CA). This bioassay compared several commercially available potting soil products with Agra Marketing's proprietary biochar blend using peat-free renewable fiber, biochar, and other locally sourced components from the Pacific Northwest. Agra Marketing is supplying this blend to large nurseries that are happy to have this sustainable, low carbon footprint potting media.

Germination and Growth Test Report for Agra Marketing Group

Test soils

1—Treatment 1: Potting soil A
2—Treatment 2: Potting soil B
3—Treatment 3: Potting soil C
4—Treatment 4: Agra Marketing biochar blend
5—Control: Sunshine Mix #4

Methodology

- Four replicates of each treatment.
- Pots are 4-inch round pots with individual humidity domes.
- 325 milliliters of damp soil in each pot.
- Eleven cucumber seeds planted in each pot with 50 milliliters plain soil on top.
- Seeds: Marketmore 76 cucumber. Johnny's Seeds, lot 72486.
- Each pot watered to saturation with pH-neutral tap water.
- Humidity domes attached.
- Pots randomized in two blocks on two greenhouse trays and placed in growth chamber on October 7, 2021.
- Lighting schedule: on 5 am to 7 pm daily.
- Pots removed from growth chamber on October 20, 2021.
- Counted germination numbers and secondary leaves.
- Plants cut at root level and aboveground biomass was harvested and placed in drying oven. After drying, biomass of each treatment weighed with scale (MyWeigh Balance 601).

Results

Table A.1. Germination and Growth Test Report for Agra Marketing Group—Results

Treatment	Germination (%)	Secondary Leaves (count)	Biomass (grams)
1	95	33	1.25
2	85	22	0.98
3	90	32	1.2
4	100	42	2.44
5	90	6	0.77

Figure A1.6. Photograph of test pots sorted by treatment.

APPENDIX B

Units and Conversion Factors

A note on units used in this book: I tend to switch constantly between metric and American units. American units are the same as the British imperial system of units except for the volume of liquids. A British pint is 20 percent larger than an American pint, which I guess is so they can drink more beer.

I use American units a lot, because it is how I think and because of where I live and the industries in which I work. In the United States, people tend to mostly use American units in cooking, agriculture, and forestry. As an engineer, I also use metric units, which are more universal. I have become adept at switching between systems, out of necessity, but I also like the flexibility of the different systems for another reason. Metric units are good when measuring very large and very small quantities that we do not encounter in person, but American and imperial units make more sense for working with systems that are human-sized because they are based on the dimensions of a human body.

While a foot is bigger than my foot, it was about the size of my father's foot. My mother would measure a yard of cloth by holding one end of the cloth to her nose and the other end in her outstretched arm to make 36 inches. An inch is just about the same size as the first joint of my thumb.

One quart and one liter are pretty close and interchangeable for the kind of imprecise "cooking" that I do. Thirty centimeters is about the same as one foot and 10 centimeters is about 4 inches, or the width of my hand at its widest point. A 5-gallon bucket is a basic unit of measurement for me, as is a 55-gallon drum. If I need to visualize a cubic yard, which is 202 gallons, I think of what can fit into four drums not quite full to the top.

But when I want to formulate a precise recipe for a spray, or measure treatment volumes for a pot trial, milliliters are my unit of choice. Fractions of an ounce are too hard to measure. A graduated cylinder with a milliliter scale is perfect for the job.

Here are some handy conversion factors:

1 teaspoon = 4.9 milliliters (5 milliliters is close enough)
1 tablespoon = 14.8 milliliters (15 milliliters is close enough)
¼ cup = 59.1 milliliters (60 milliliters is close enough)
1 quart = 0.95 liter
1 cubic foot = 0.3 cubic meters = 7.5 gallons = 1.5 five-gallon buckets
5 gallons = ⅔ cubic foot = 20 quarts = 19 liters
1 cubic yard = 0.75 cubic meters = 202 gallons (4 not-quite-full 55-gallon drums) = 40.5 five-gallon buckets
200 liters = 52.8 gallons
1,000 liters = 1 cubic meter
1 cubic meter = 1.3 cubic yards = 264 gallons (5 almost-full 55-gallon drums)
1 cord of wood (4 feet × 4 feet × 8 feet) = 4.75 cubic yards envelope volume, 3.25 cubic yards solid wood (subtracting air spaces)
1 kilogram = 2.2 pounds
1 metric tonne = 2,204 pounds
1 US short ton = 2,000 pounds
1 tonne of biochar carbon = 3.67 tonnes of carbon dioxide (3.67 is the conversion factor that adds two molecules of oxygen to each molecule of carbon)

Glossary

Active carbon: Soil carbon that is easily oxidized and readily available as a carbon and energy source for microbes.

Activated charcoal: Charcoal that is treated with hot gas, steam, or chemicals to expand the internal surface area of the material, increasing its ability to adsorb target chemicals.

Absorption: The process by which one substance, such as a solid or liquid, takes up another substance, such as a liquid or gas, by filling pores, voids, or spaces in the first substance.

Adiabatic flame temperature: The maximum temperature reached by a flame under ideal conditions.

Adsorption: The adhesion in an extremely thin layer of molecules to the internal and external surfaces of a substance.

Aerobic: A process in combustion or in microbial respiration that takes place with free access to air or oxygen.

Air curtain incinerator: An insulated box with a blower that projects a high-velocity curtain of air into the box for the purpose of controlling air emissions during burning.

Anaerobic: A process in combustion or in microbial respiration that takes place with restricted access to air or oxygen.

Anion: A negatively charged ion. In an electrolyte, as in a battery, an anion is attracted to the positively charged electrode, called the anode.

Anthrosol: A soil created or modified by humans, as in the terra preta soils of the Amazon or the plaggen soils of northern Europe.

Ash: The inorganic matter or mineral residue of biomass that remains when combusted in the presence of excess air.

Aromatic carbon: Carbon that is in the form of six-sided carbon molecules linked together in sheets or chains.

Autothermal process: A thermochemical reaction that pyrolyzes organic materials using only the heat produced by the reaction itself.

Benzene ring: A benzene molecule, composed of six carbon atoms joined in a hexagonal ring structure with one hydrogen atom attached to each atom of carbon. This is the simplest aromatic hydrocarbon based on a hexagonal carbon ring.

Biochar: A solid material obtained from thermochemical conversion of biomass in an oxygen-limited environment that is suitable for use in soils or for carbon sequestration.

Glossary

Biochar mineral complex: A form of organo-mineral complex based on biochar as the organic matter that is bound to mineral particles to create stable soil particles.

Biochar physicochemical properties: Those physical and chemical properties of biochar that affect the uses of biochar in soils and the environment.

Biochar quality: Biochar quality is assessed according to the purpose of the biochar use. In the case of biochar used for carbon sequestration in soil, biochar quality is determined by the recalcitrance of the carbon in the biochar.

Biomass: The biodegradable fraction of biological material derived from recently living beings, usually plants.

Bokashi: A grain or other carbohydrate fermented with lactic acid bacteria and other beneficial microorganisms, often made with EM-1 inoculant.

Carbon drawdown: Sequestration of carbon dioxide in the biosphere through deliberate farming, grazing, reforestation, and waste management practices.

Carbonization: The conversion of an organic substance into a stable carbon residue through pyrolysis.

Cation: A positively charged ion. In an electrolyte, as in a battery, a cation is attracted to the negatively charged electrode, called the cathode.

Charcoal: A solid material obtained from thermochemical conversion of biomass in an oxygen-limited environment. All biochar is charcoal, but not all charcoal is biochar.

Charged biochar: Biochar that has been treated with nutrients, beneficial microbes, or both to create a fertilizing soil amendment.

Complexation: A combination of individual atom groups, ions, or molecules to create one large ion or molecule.

Conservation burn pile: A method of open pile burning that manages the burn process to produce minimum smoke and ash and maximum biochar.

De-volatilization: Release of hydrocarbon gases from biomass through heat or drying.

Dissolved organic carbon (DOC): Organic matter that can pass through a filter that is between 0.22 and 0.7 microns. DOC is an important nutrient source for microbes.

Diffusion flame: A flame where the fuel and oxidizer are not premixed and come together by physical mixing and molecular diffusion of the fuel and oxidizer.

Electron acceptor: A chemical species that accepts electrons transferred to it from another compound. It is an oxidizing agent that, by virtue of its accepting electrons, is itself reduced in the process.

Electron donor: A chemical species that donates electrons to another compound. It is a reducing agent that, by virtue of its donating electrons, is itself oxidized in the process.

EM-1: Stands for Effective Microorganisms, a consortium of microorganisms that work together in a self-propagating culture. EM-1 is available commercially from several suppliers.

Exudate: In plants, root exudates are substances that plants use to attract beneficial fungi and microbes and repel pathogenic species.

Feedstock: The material undergoing the thermochemical process to create biochar. Feedstock material for biochar is biomass from plants and other organisms.

Fixed carbon: Another term for *recalcitrant carbon*.

Flame: Burning gas that emits heat and light.

Flame-cap kiln: A carbonizer that uses a cap of flame both to transfer heat to the feedstock and to prevent oxidation of the finished biochar by consuming all of the available oxygen.

Flame carbonization: A process for carbonizing biomass in the presence of an oxidizing flame that relies on stopping the combustion process before the material turns to ash.

Forest residues: Woody material not harvested and removed from logging sites in commercial operations, as well as material resulting from forest management operations such as fuels-reduction thinning and removal of dead and dying trees.

Fulvic acid: Low molecular weight acidic polymer that can be extracted from soil humus.

Functional group: In organic chemistry, a portion of a molecule that causes a characteristic chemical reaction. The same functional group will undergo the same or similar chemical reactions regardless of the structure of the rest of the molecule.

Fused carbon rings: Another term for *aromatic carbon*.

Gasifier: A thermochemical reactor optimized to produce gas from a fuel. The emphasis is on producing gas rather than char. In contrast, a carbonizer produces gas while optimizing char production.

Greenhouse gas (GHG): Any of the gases whose absorption of solar radiation is responsible for the greenhouse effect, including carbon dioxide, methane, nitrous oxide, and the fluorocarbons.

Graphene: A single layer of aromatic carbon molecules linked together in a sheet.

Graphite: Mineral composed of many layers of graphene sheets.

Heat treatment temperature (HTT): The temperature at which a biomass feedstock material is processed during a thermochemical conversion process to create biochar.

Humic acid: Medium molecular weight acidic polymer that can be extracted from soil humus, with more aromatic carbon in its structure than fulvic acid.

Humin: High molecular weight organic polymers that are not soluble in water, acid, or alkaline extractions.

Humate: Mineral salts of humic acids.

Humus: A dark-colored constituent of soil composed of large, stable organic complexes that include a large proportion of aromatic carbon rings.

Incineration burn pile: A pile of organic waste that is burned in the open air for purposes of incineration and waste disposal.

Indigenous microorganisms: Beneficial microbes native to a local environment than can be cultured and applied to agricultural soils for soil health and plant growth benefits. Usually includes fungi, yeasts, and bacteria.

Inorganic carbon: In soil, the carbonates such as calcium carbonate.

Inter-particle space: The spaces in biochar soil mixes where water can be held between biochar particles and other soil particles.

Intra-particle space: The porous spaces that can hold water within a biochar particle.

Labile carbon: Carbon that can be more easily degraded by microorganisms than fixed carbon. Labile carbon includes active carbon.

Lactic acid bacteria (LAB): A broad group of bacteria used in fermentation and food preservation by creating an acidic environment. LAB are ubiquitous in natural environments.

Laminar flame: Flames that flow smoothly with limited turbulent mixing of fuel and air. Mixing is by molecular diffusion of the reactants.

Life cycle assessment (LCA): Method for the environmental assessment of products covering their life cycle from raw material extraction to waste treatment (cradle-to-grave) or to the regeneration of the raw material (cradle-to-cradle).

Ligand: An ion or a molecule with a functional group that binds to a metal atom.

Methanogens: Archaea that produce methane as a waste product of metabolism in low-oxygen conditions.

Methanotrophs: Bacteria and archaea that oxidize methane.

Mineral carbon: Mineral carbon refers to carbon solids such as diamond and graphite, as well as the gases of carbon (CO_2, CO).

Mineralization: The process, usually mediated by microbes, that converts an organic compound into an inorganic compound in soil.

Mobile matter: Like volatile matter and labile carbon, a term for the fraction of biochar that can be degraded by microbes and converted to carbon dioxide and other greenhouse gases.

Mollisols: Black earth soils with a high content of black carbon formed by grassland fires that are among the most productive agricultural soils. Also known as chernozems.

Nitrification: The biological process that converts ammonia to nitrates that are easily used by plants.

Nitrogen immobilization: When microbes take up free nitrogen in soil as part of their metabolism of organic carbon, the nitrogen is immobilized in their bodies until they die and degrade.

Nitrogen mineralization: When microbes decompose organic, nitrogen-containing compounds from manure and other sources, converting them to ammonium.

Open burn pile: An ordinary burn pile or campfire that does not take place inside a container of any kind.

Organic carbon: The carbon in compounds that are found in living or once-living things. Those compounds include hydrogen and oxygen in the form of hydrocarbons and carbohydrates. The organic carbon fraction does not include inorganic carbonates such as calcium and magnesium carbonates.

pH: Potential hydrogen, a measure of the hydrogen-ion concentration in a solution that determines acidity or alkalinity.

Pfeiffer's circular chromatography (PCC): A chromatography method that applies an alkaline extraction of an organic substance to circular paper filters treated with a photosensitive silver nitrate solution for the purpose of creating a visual pattern that indicates certain qualities of the material.

Phospholipid fatty acids (PLFA): Fatty acids found in the membranes of living organisms and are used to identify classes of bacteria, fungi, and other soil life.

Pi bonds: A type of diffuse atomic bond formed by the overlap of the p orbitals on adjacent atoms.

Plaggen soils: Created in parts of northern Europe when farmers cut turf (Dutch, *plaggen*) to use as bedding for cattle, and then transferred manure-enriched plaggen to fields.

Plant secondary metabolite (PSM): Small molecular substances made by plants and that work on the plant itself or act as antibiotics, insect repellents, attractants, and other purposes. PSM are the basis of many herbal medicines.

Polyaromatic hydrocarbon (PAH): A family of compounds based on two or more benzene rings formed as products of combustion. Out of hundreds of different PAH compounds, only a few are considered to be highly toxic and of regulatory concern.

Porosity: A measure of the empty space or voids within a material, expressed as the volume of voids over the total volume.

Proximate analysis: Fractional analysis of an organic substance to determine the content of moisture, ash, fixed carbon, and volatile matter.

Purple non-sulfur photosynthetic bacteria (PNSB): A group of phototrophic bacteria that degrade organic matter and produce hydrogen and plant-growth-promoting substances.

Pyrolysis: The thermochemical decomposition of organic material in an oxygen-limited environment into char, gas, liquids, and ash.

Pyroligneous acid: An acidic liquid produced by the destructive distillation of wood or other biomass.

Raw biochar: Biochar straight from the kiln with no other additives but water.

Recalcitrant carbon: The fraction of soil organic matter that is resistant to microbial decomposition.

Redox reactions: These always occur in pairs and involve the transfer of electrons between chemical species where one is oxidized (loses an electron) and one is reduced (gains an electron). The species that gains an electron is called "reduced" because its charge becomes more negative. Also known as oxidation-reduction reactions.

Residence time: The time a feedstock is held within a consistent temperature range in a given thermochemical process.

Retort: A closed vessel with a gas outlet and that can be heated to cause the substances within it to pyrolyze, separating gases from solids.

Sequestered carbon: The recalcitrant carbon content of biochar that is added to soil.

Stable carbon: Another term to refer to recalcitrant or sequestered carbon that is not easily degradable by microbes.

Specific surface area: The surface area of the pores inside biochar, reported as area per unit of mass of the sample—for instance, as square meters per gram.

Terra preta: Deep, fertile, black anthrosol created by Amazonians who applied charcoal combined with organic waste to crop fields over thousands of years. From Portuguese, meaning "dark earth."

Terra preta sanitation: Sanitation system for human waste that includes urine diversion, addition of a charcoal mixture, and lactic acid fermentation with subsequent vermicomposting. Developed by researchers at the Technical University of Hamburg as an extrapolation from the practices of Amazonians who created terra preta.

Thermochemical conversion: Thermochemical conversion of biomass includes combustion, pyrolysis, and gasification. A thermochemical reaction may convert wood or other fuel into products such as heat, gas, oil, distilled liquids, char, and ash.

TLUD: Top-lit updraft gasifier. A type of simple gasifier that can create usable heat and biochar.

Torrefaction: A form of mild pyrolysis that takes place at temperatures of 200 to 350°C. Coffee roasting is an example of torrefaction.

Turbostratic carbon: Carbon structures that are somewhere between amorphous (unstructured) and graphitic (layered sheets) carbon.

Turbulent flame: The degree of turbulence in a fluid system characterized by the balance of energy and friction in the system. In a diffusion flame, turbulence helps with mixing of fuel and air by adding physical movement of fuel and air to bring the reactants closer together.

Ultimate analysis: Chemical analysis of an organic substance broken down into the elements of carbon, hydrogen, oxygen, nitrogen, and sulfur.

Volatile organic compounds (VOC): Emissions from combustion that are harmful to health. VOC emissions from burning wood can include acetic acid, furfural, toluene, terpenes, PAH, and many other compounds.

Volatile matter: In biomass, volatile matter is anything that burns, excluding mineral ash. In biochar, volatile matter refers to the labile carbon that is degradable by microbes.

Wood vinegar: Another name for *pyroligneous acid*.

Notes

Chapter 1. Biochar in History

1. Johannes Lehmann and Stephen Joseph, eds., *Biochar for Environmental Management: Science, Technology and Implementation* (Routledge, 2015).
2. Eduardo G. Neves et al., "Historical and Socio-Cultural Origins of Amazonian Dark Earth," in *Amazonian Dark Earths: Origin Properties Management*, eds. J. Lehmann et al. (Dordrecht: Springer, 2003), 29–50, https://doi.org/10.1007/1-4020-2597-1.
3. Charles C. Mann, *1491: New Revelations of the Americas Before Columbus* (New York: Vintage, 2006).
4. Bruno Glaser and Jago Jonathan Birk, "State of the Scientific Knowledge on Properties and Genesis of Anthropogenic Dark Earths in Central Amazonia (*terra preta de Índio*)," *Geochimica et Cosmochimica Acta* 82 (2012): 39–51.
5. Mark B. Bush and Miles R. Silman, "Amazonian Exploitation Revisited: Ecological Asymmetry and the Policy Pendulum," *Frontiers in Ecology and the Environment* 5, no. 9 (2007): 457–65.
6. Douglas Sheil et al., "Do Anthropogenic Dark Earths Occur in the Interior of Borneo? Some Initial Observations from East Kalimantan," *Forests* 3, no. 2 (2012): 207–29.
7. James Fraser, James Fairhead, and Melissa Leach, "Green Grabs and Biochar: Revaluing African Soils and Farming in the New Carbon Economy," *Journal of Peasant Studies* 39, no. 2 (2012): 285–307.
8. N. Sai Bhaskar Reddy, "Soils Near Batibo, Cameroon," Terra Preta Info, February 14, 2009, https://e-terrapreta.blogspot.com/2009/02/soils-near-batibo-cameroon.html.
9. José Ramón Olarieta et al., "'Formiguers,' a Historical System of Soil Fertilization (and Biochar Production?)," *Agriculture, Ecosystems & Environment* 140, no. 1–2 (2011): 27–33.
10. Franklin Hiram King, *Farmers of Forty Centuries: Organic Farming in China, Korea, and Japan*, Dover Edition (Madison, WI: 1911; Mineola, NY: Dover, 2004).
11. John Paull, "The Making of an Agricultural Classic: Farmers of Forty Centuries; or Permanent Agriculture in China, Korea and Japan, 1911–2011," *Agricultural Sciences* 2, no. 3 (2011): 175.
12. Franklin Hiram King, *Farmers of Forty Centuries*, 193.
13. Franklin Hiram King, *Farmers of Forty Centuries*, 194.
14. Franklin Hiram King, *Farmers of Forty Centuries*, 397.
15. Franklin Hiram King, *Farmers of Forty Centuries*, 193.
16. Franklin Hiram King, *Farmers of Forty Centuries*, 197.

17. Makoto Ogawa and Yasuyuki Okimori, “Pioneering Works in Biochar Research, Japan,” *Soil Research* 48, no. 7 (2010): 489–500.
18. Charles C. Mann, “Our Good Earth: The Future Rests on the Soil Beneath Our Feet. Can We Save It?” *National Geographic* 214, no. 3 (2008): 80–106.
19. Małgorzata Oleszkiewicz-Peralba, *The Black Madonna in Latin America and Europe* (Albuquerque: University of New Mexico Press, 2009).
20. William H. Brock, *Justus von Liebig: The Chemical Gatekeeper* (Cambridge: Cambridge University Press, 2002), 257.
21. Freiherr von Justus Liebig, *Liebig's Complete Works on Chemistry* (Philadelphia: T.B. Peterson,1852), 65–66.
22. Freiherr von Justus Liebig, *Liebig's Complete Works*, 84–86.
23. “Charcoal as a Fertilizer,” *The New Jersey Farmer* 2, no. 1 (September 1856): 75.
24. “Manures,” in *Transactions of the N.Y. State Agricultural Society, with an Abstract of the Proceedings of the County Agricultural Societies*, vol. 12 (Albany: 1853), 867.
25. John Chalmers Morton, *A Cyclopedia of Agriculture: Practical and Scientific*, vol. 1 (Blackie and Son, 1855), 431.
26. “Peat Charcoal,” *American Journal of Agriculture and Science*, 5–6 (1847).
27. Freiherr von Justus Liebig, *Liebig's Complete Works*, 14.
28. Freiherr von Justus Liebig, *Liebig's Complete Works*, 40–41.
29. Local Government Board, Her Majesty's Stationery Office, London, *Sewage Disposal: Report of a Committee*, 1876.
30. Freiherr von Justus Liebig, *Liebig's Complete Works*, 66.
31. Freiherr von Justus Liebig, *Liebig's Complete Works*, 66.
32. Freiherr von Justus Liebig, *Liebig's Complete Works*, 66.
33. William H. Brock, *Justus von Liebig*, 271.
34. Jasper W. Rogers, *Facts and Fallacies of the Sewerage System of London and Other Large Towns* (Atchley, 1857).
35. Jasper W. Rogers, *The Potato Truck System of Ireland: The Main Cause of Her Periodical Famines and of the Non-Payment of Her Rents* (J. Ridgway, 1847).
36. Jasper W. Rogers, *Facts and Fallacies of the Sewerage System*.
37. “Drying Peat to Make Charcoal at Derrymullen, Kildare, Ireland,” *Illustrated London News* XVII, no. 448 (September 28, 1850), 264.
38. House of Commons, Great Britain, *Parliamentary Papers*, vol. 38, March 20, 1877, 104.
39. William Ashwell Shenstone, *Justus von Liebig: His Life and Work (1803–1873)* (London: Cassell, 1895).

Chapter 2. What Is Biochar and How Does It Work?

1. Johannes Lehmann and Markus Kleber, “The Contentious Nature of Soil Organic Matter,” *Nature* 528, no. 7580 (2015): 60–68.
2. Richard Fortey, *Life: A Natural History of the First Four Billion Years of Life on Earth* (New York: Vintage, 2011).
3. Edward D. Goldberg, *Black Carbon in the Environment: Properties and Distribution* (New York: Wiley and Sons, 1985).
4. Sylvia M. Old, “Microclimate, Fire, and Plant Production in an Illinois Prairie,” *Ecological Monographs* 39, no. 4 (1969): 355–84.

5. Jan O. Skjemstad et al., "Charcoal Carbon in US Agricultural Soils," *Soil Science Society of America Journal* 66, no. 4 (2002): 1249–55; Bruno Glaser and Wulf Amelung, "Pyrogenic Carbon in Native Grassland Soils along a Climosequence in North America," *Global Biogeochemical Cycles* 17, no. 2 (2003).
6. J-D Mao et al., "Abundant and Stable Char Residues in Soils: Implications for Soil Fertility and Carbon Sequestration," *Environmental Science & Technology* 46, no. 17 (2012): 9571–76.
7. Marco Keiluweit et al., "Dynamic Molecular Structure of Plant Biomass-Derived Black Carbon (Biochar)," *Environmental Science & Technology* 44, no. 4 (2010): 1247–53.
8. Derek R. Lovley et al., "Humic Substances as Electron Acceptors for Microbial Respiration," *Nature* 382, no. 6590 (1996): 445–48.
9. Andreas Kappler et al., "Biochar as an Electron Shuttle between Bacteria and Fe (III) Minerals." *Environmental Science & Technology Letters* 1, no. 8 (2014): 339–44.
10. Catherine E. Brewer et al., "New Approaches to Measuring Biochar Density and Porosity," *Biomass and Bioenergy* 66 (2014): 176–85.
11. A. Downie, A. Crosky, and P. Munroe, "Physical Properties of Biochar," in *Biochar for Environmental Management: Science and Technology*, eds. Johannes Lehman and Stephen M. Joseph (London: Routledge, 2009), 448.
12. Laura Klüpfel et al., "Redox Properties of Plant Biomass-Derived Black Carbon (Biochar)," *Environmental Science & Technology* 48, no. 10 (2014): 5601–11.
13. M. H. B. Hayes, "Relationships between Biochar and Soil Humic Substances," in *Functions of Natural Organic Matter in Changing Environment* (Springer Netherlands, 2013), 959–63.
14. Biqing Liang et al., "Black Carbon Affects the Cycling of Non-Black Carbon in Soil," *Organic Geochemistry* 41, no. 2 (2010): 206–13.
15. Humberto Blanco—Canqui et al., "Soil Carbon Increased by Twice the Amount of Biochar Carbon Applied after 6 Years: Field Evidence of Negative Priming," *GCB Bioenergy* 12, no. 4 (2020): 240–51.
16. Zhe Weng et al., "Microspectroscopic Visualization of How Biochar Lifts the Soil Organic Carbon Ceiling," *Nature Communications* 13, no. 1 (2022): 5177.
17. Minh Ky Nguyen et al., "Evaluate the Role of Biochar During the Organic Waste Composting Process: A Critical Review," *Chemosphere* 299 (2022): 134488.
18. Thomas H. DeLuca et al., "Biochar Effects on Soil Nutrient Transformations," in *Biochar for Environmental Management: Science, Technology and Implementation*, vol. 2, eds. Johannes Lehmann and Stephen Joseph (Routledge, 2015), 421–54.
19. Keiji Jindo et al., "Chemical and Biochemical Characterisation of Biochar-Blended Composts Prepared from Poultry Manure," *Bioresource Technology* 110 (2012): 396–404.
20. Jining Zhang et al., "The Use of Biochar-Amended Composting to Improve the Humification and Degradation of Sewage Sludge," *Bioresource Technology* 168 (2014): 252–58.
21. Hardy Schulz, Gerald Dunst, and Bruno Glaser, "Positive Effects of Composted Biochar on Plant Growth and Soil Fertility," *Agronomy for Sustainable Development* 33 (2013): 817–27.
22. Daniel Fischer and Bruno Glaser, "Synergisms between Compost and Biochar for Sustainable Soil Amelioration," *Management of Organic Waste* 1 (2012): 167–98.

Notes

Chapter 3. How to Make Your Own Biochar

1. Ya Inoue, Kunitoyo Mogi, and S. J. A. K. Yoshizawa, "Properties of Cinders from Red Pine, Black Locust and Henon Bamboo," proceedings from Asia Pacific Biochar Conference, Kyoto, 2011.

Chapter 4. Working the Flame

1. P. H. Thomas, R. Baldwin, and A. J. M. Heselden, "Buoyant Diffusion Flames: Some Measurements of Air Entrainment, Heat Transfer, and Flame Merging," *Symposium (International) on Combustion* 10, no. 1 (January 1965): 983–96.
2. Christa Roth, *Micro-Gasification Cooking with Gas from Dry Biomass: An Introduction to Concepts and Applications of Wood-Gas Burning Technologies for Cooking* (GIZ Hera, 2011).
3. R. Scharler et al., "Emission Minimization of a Top-Lit Updraft Gasifier Cookstove Based on Experiments and Detailed CFD Analyses," *Energy Conversion and Management* 247 (2021), 114755.
4. Kelpie Wilson and Tom Miles, "Combined Heat and Biochar Technology Assessment for a Composting Operation," US Biochar Initiative, August 2020.

Chapter 5. Biochar Properties

1. Marco Keiluweit et al., "Dynamic Molecular Structure of Plant Biomass-Derived Black Carbon (Biochar)," *Environmental Science & Technology* 44, no. 4 (2010): 1247–53.
2. Paul Taylor, ed., *The Biochar Revolution* (Mt. Evelyn, Victoria: Global Publishing Group, 2010).
3. Catherine Brewer and Jonah Levine, "Weight or Volume for Handling Biochar and Biomass?", *The Biochar Journal* (2015), https://biochar-journal.org/en/ct/71.
4. Jihua Zhai et al., "New Insights into Biomass Combustion Ash Categorisation: A Phylogenetic Analysis," *Fuel* 287 (March 2021), 119469, https://doi.org/10.1016/j.fuel.2020.119469.
5. Kelpie Wilson and Debra Reed, "Implications and Risks of Potential Dioxin Presence in Biochar," International Biochar Initiative, April 2012.
6. G. Cornelissen et al., "Emissions and Char Quality of Flame-Curtain 'Kon Tiki' Kilns for Farmer-Scale Charcoal/Biochar Production," *PloS One* 11, no. 5 (May 2016): e0154617.
7. R. B. Fidel, D. A. Laird, and K. A. Spokas, "Sorption of Ammonium and Nitrate to Biochars Is Electrostatic and Ph-Dependent," *Scientific Reports* 8, no. 1 (2018): 17627.
8. S. Adhikari et al., "Evaluating Fundamental Biochar Properties in Relation to Water Holding Capacity," *Chemosphere* 328 (2023): 13862.
9. John McDonald-Wharry, "2013–2014 Survey of Chars Using Raman Spectroscopy," *C Journal of Carbon Research* 7, no. 3 (2021): 63.
10. Francesco Restuccia et al., "Quantifying Self-Heating Ignition of Biochar as a Function of Feedstock and the Pyrolysis Reactor Temperature," *Fuel* 236 (2019): 201–13.

Chapter 6. Transforming Smoke into Useful Chemicals

1. Office of Her Royal Highness Princess Maha Chakri, "Wood Vinegar Pamphlet," Bangkok, Thailand.

2. J. Mu, T. Uehara, and T. Furuno, "Effect of Bamboo Vinegar on Regulation of Germination and Radicle Growth of Seed Plants," *Journal of Wood Science* 49, no. 3 (June 2003): 262–70, http://doi.org/10.1007/ s10086-002-0472-z.
3. M. Hagner, "Potential of the Slow Pyrolysis Products Birch Tar Oil, Wood Vinegar and Biochar in Sustainable Plant Protection: Pesticidal Effects, Soil Improvement and Environmental Risks," (PhD diss., University of Helsinki, 2013).
4. J. Mu et al., "Preliminary Study of Application Effect of Bamboo Vinegar on Vegetable Growth," *Forestry Studies in China* 8, no. 3 (2006): 43–47, https://doi.org/10.1007 / s11632-006-0023-6.
5. B. Zulkarami et al., "Effect of Pyroligneous Acid on Growth, Yield and Quality Improvement of Rockmelon in Soilless Culture," *Australian Journal of Crop Science* 5, no. 12 (2011): 1508–14.
6. K. Zhang et al., "Germination and Growth Performance of Water-Saving and Drought-Resistant Rice Enhanced by Seed Treatment with Wood Vinegar and Biochar Under Dry Direct-Seeded System," *Agronomy* 12, no. 5 (2022): 1223.
7. J. Cheon-Soon et al., "Effect of Wood Vinegar and Charcoal on Growth and Quality of Sweet Pepper," *Korean Journal of Horticultural Science & Technology* 24, no. 2 (2006): 177–80.
8. K. Zhu et al., "Wood Vinegar Impact on the Growth and Low-Temperature Tolerance of Rapeseed Seedlings," *Agronomy* 12, no. 10 (2022): 2453.
9. J. Ma et al., "Wood Vinegar Induces Salinity Tolerance by Alleviating Oxidative Damages and Protecting Photosystem II in Rapeseed Cultivars," *Industrial Crops and Products* 189 (2022): 115763.
10. K. H. Jung, "Growth Inhibition Effect of Pyroligneous Acid on Pathogenic Fungus, *Alternaria Mali*, the Agent of Alternaria Blotch of Apple," *Biotechnology and Bioprocess Engineering* 12 (2007): 318–22.
11. T. Gao et al., "Wheat Straw Vinegar: A More Cost-Effective Solution than Chemical Fungicides for Sustainable Wheat Plant Protection," *Science of the Total Environment* 725 (2020): 138359.
12. G. Indriati and M. Puspitasari, "The Potency of Coconut Shell Wood Vinegar and Essential Oils as Botanical Insecticides to Control *Hypothenemus hampei* (Coleoptera: Curculionidae)," *IOP Conference Series: Earth and Environmental Science* 762, no. 1 (May 2021): 012057.
13. X. Guo et al., "Influence of Pyroligneous Acid on Fermentation Parameters, CO_2 Production and Bacterial Communities of Rice Straw and Stylo Silage," *Frontiers in Microbiology* 12 (2021): 701434.
14. E. J. Calabrese and E. Agathokleous, "Smoke-Water Commonly Induces Hormetic Dose Responses in Plants," *Science of The Total Environment* 765 (2021): 142776.
15. Hirowaka Tsuyoshi, *The Uses of Wood Vinegar and Charcoal in Agriculture: How to Improve Crop Quality While Reducing Dependence on Agricultural Chemicals* (Pak Chong, Thailand: Asia Regional Cookstove Program, 1994).
16. Masaki Yokomori, *Farmers in Benguet Practice Savers Technology* (Japan International Cooperation Agency, Safe Vegetable Promotion Project in Benguet: 2011).

17. Boonsong Thansrithong and Patrick Trail, "An Innovative 'Do-All' Biochar Burner Design," ECHO Asia Impact Center, Chiang Mai, Thailand, published September 1, 2021, http://edn.link/6x6qfx.

Chapter 7. Ashes and the Value of Minerals in Biochar

1. Alicia Reynolds et al., "Effect of Clay and Iron Sulphate on Volatile and Water-Extractable Organic Compounds in Bamboo Biochars," *Journal of Analytical and Applied Pyrolysis* 133 (2018): 22–29.
2. Stephen Joseph et al., "Effects of Enriched Biochars Containing Magnetic Iron Nanoparticles on Mycorrhizal Colonisation, Plant Growth, Nutrient Uptake and Soil Quality Improvement," *Pedosphere* 25, no. 5 (2015): 749–60.
3. Jun Ye et al., "A Combination of Biochar–Mineral Complexes and Compost Improves Soil Bacterial Processes, Soil Quality, and Plant Properties," *Frontiers in Microbiology* 7 (2016): 372.
4. Peng Zhang et al., "Functional Biochar and Its Balanced Design," *ACS Environmental Au* 2, no. 2 (2021): 115–27.
5. S. L. Dana, *A Muck Manual for Farmers* (New York: C. M. Saxton & Company, 1856).
6. A. Hart et al., "Value-Added Materials Recovered from Waste Bone Biomass: Technologies and Applications," *RSC advances* 12, no. 34 (2022): 22302–30.
7. Kurt Pennell, "Specific Surface Area," *Reference Module in Earth Systems and Environmental Sciences* (Elsevier, 2016), https://doi.org/10.1016/B978-0-12-409548-9.09583-X.
8. P. Zhang et al., "Functional Biochar and Its Balanced Design," *ACS Environmental Au* 2, no. 2 (2021): 115–27.
9. Josh Kearns, *A Field Guide to Biochar Water Treatment*, joshkearns.substack.com.
10. S. S. Alkurdi et al., "Bone Char as a Green Sorbent for Removing Health Threatening Fluoride from Drinking Water," *Environment International* 127 (2019): 704–19.
11. Margaret O. Thompson and Joshua P. Kearns, "Modeling and Experimental Approaches for Determining Fluoride Diffusion Kinetics in Bone Char Sorbent and Prediction of Packed-Bed Groundwater Defluoridator Performance," *Water Research X* 12 (2021): 100108.
12. Josh Kearns, "Book-in-Progress: Chapter 4—Summary of Research on Biochar Adsorption of Chemical Pollutants: How Semi-Aerobic Gasification Generates Enhanced Biochar Adsorbent," *A Field Guide to Biochar Water Treatment*, January 9, 2021, joshkearns.substack.com.
13. J. P. Kearns et al., "High Temperature Co-Pyrolysis Thermal Air Activation Enhances Biochar Adsorption of Herbicides from Surface Water," *Environmental Engineering Science* 36, no. 6 (2019).
14. A. L. Gao and Y. Wan, "Iron Modified Biochar Enables Recovery and Recycling of Phosphorus from Wastewater Through Column Filters and Flow Reactors," *Chemosphere* 313 (2023): 137434; L. Hao et al., "A Critical Review on Arsenic Removal from Water Using Iron-Based Adsorbents," *RSC Advances* 8, no. 69 (2018): 39545–60.
15. Ian Donaldson, "How to Make Copperas (Iron Sulfate) from Pyrites," accessed August 8, 2023, https://www.juliantrubin.com/encyclopedia/chemistry/copperas.html.
16. C. J. Mitchell, A. D. Jayakaran, and J. K. McIntyre, "Biochar and Fungi as Bioretention Amendments for Bacteria and PAH Removal from Stormwater," *Journal of Environmental Management* 327 (2023): 116915.

17. Josh Kearns, "The End of Global Development as We Know It," Resilience.org, December 4, 2014, https://www.resilience.org/stories/2014-12-04/the-end-of-global-development-as-we-know-it/.
18. N. Redvers et al., "Uranium Exposure in American Indian Communities: Health, Policy, and the Way Forward," *Environmental Health Perspectives* 129, no. 3 (2021): 035002.

Chapter 8. A Modern Muck Manual

1. C. W. Ritz et al., "Evaluation of Surface-Applied Char on the Reduction of Ammonia Volatilization from Broiler Litter," *Journal of Applied Poultry Research* 20, no. 2 (2011): 240–45.
2. Xiuzhang Wang et al., "Greenhouse Gas Reduction and Nitrogen Conservation During Manure Composting by Combining Biochar with Wood Vinegar," *Journal of Environmental Management* 324 (2022): 116349.
3. Cen Zhang et al., "Using Poultry Litter Derived Biochar as Litter Amendment to Control Ammonia Emissions," in *2016 ASABE Annual International Meeting* (American Society of Agricultural and Biological Engineers, 2016).
4. M. P. Bernal, J. A. Alburquerque, and Raúl Moral, "Composting of Animal Manures and Chemical Criteria for Compost Maturity Assessment. A Review," *Bioresource Technology* 100, no. 22 (2009): 5444–53.
5. Keiji Jindo et al., "Biochar Influences the Microbial Community Structure During Manure Composting with Agricultural Wastes," *Science of the Total Environment* 416 (2012): 476–81.
6. Michael W. DuPonte and David Fischer, "The Natural Farming Concept: A New Economical Waste Management System for Small Family Swine Farms in Hawai'i," *LM*-23, University of Hawai'i, College of Tropical Agriculture and Human Resources, 2012.
7. Hans-Peter Schmidt et al., "The Use of Biochar in Animal Feeding," *PeerJ* 7 (2019): e7373.
8. T. Higa and J. F. Parr, *Beneficial and Effective Microorganisms for a Sustainable Agriculture and Environment* (Atami, Japan: International Nature Farming Research Center, 1994).
9. Adam Footer, *Bokashi Composting* (Gabriola, Canada: New Society Publishers, 2014).
10. S. De Gisi, L. Petta, and C. Wendland, "History and Technology of Terra Preta Sanitation," *Sustainability* 6, no. 3 (2014): 1328–45.
11. Sally Brown, "Connections: How Much N And P Are in Urban Residuals?" Biocycle, August 7, 2023, https://www.biocycle.net/connections-nitrogen-and-phosphorus-in-urban-waste/.
12. Carol Steinfield, *Liquid Gold* (Gabriola, B.C.: New Society Publishers, 2007).

Chapter 9. Biochar in Food and Medicine

1. Stephen Joseph et al., "Feeding Biochar to Cows: An Innovative Solution for Improving Soil Fertility and Farm Productivity," *Pedosphere* 25, no. 5 (2015): 666–79; Lauren Celenza, "Whole-Farm Biochar System Boosts Productivity, Stores Carbon, Cuts Inputs and Emissions," *WANTFA*, Summer 2015.
2. "Quantification Protocol for Reducing Greenhouse Gas Emissions from Fed Cattle," Version 3.0, Alberta Government, February 2016, https://open.alberta.ca/dataset/ec8dce85-297c-45d2-82a6-58bbfed85a00/resource/2cd6e6a7-fd63-41c3-8a61-8f0567f90b0b/download/reducingghgemissionsfedcattle-feb25-2016.pdf.

3. T. Tseten et al., "Strategies to Mitigate Enteric Methane Emissions from Ruminant Animals," *Journal of Microbiology and Biotechnology* 32, no. 3 (March 2022): 269–77.
4. R. A. Leng, S. Inthapanya, and T. R. Preston, "Biochar Lowers Net Methane Production from Rumen Fluid in Vitro," *Livestock Research for Rural Development* 24, no. 6 (2012).
5. R. A. Leng, T. R. Preston, and S. Inthapanya, "Biochar Reduces Enteric Methane and Improves Growth and Feed Conversion in Local 'Yellow' Cattle Fed Cassava Root Chips and Fresh Cassava Foliage," *Livestock Research for Rural Development* 24, no. 11 (2012).
6. R. A. Leng, S. Inthapanya, and T. R. Preston, "Methane Production Is Reduced in an In Vitro Incubation when the Rumen Fluid Is Taken from Cattle that Previously Received Biochar in Their Diet," *Livestock Research for Rural Development* 24, no. 11 (2012).
7. A. R. Rodrigues et al., "Heads or Tails? Toward a Clear Role of Biochar as a Feed Additive on Ruminant's Methanogenesis," in *Biochar: Productive Technologies, Properties and Applications*, eds. M. Bartoli, M. Giorcelli, A. Tagliaferro, (London: IntechOpen, 2023).
8. M. H. Sirjani et al., "Effect of Adding Biochar in Diets Containing Probiotics on In Vitro Fermentation Variables, Health Indicators, Rectum Bacteria, and Blood Enzymes of Holstein Calves," *Animal Production Research* 11, no. 4 (2023): 1–19.
9. D. Van Dung et al., "Effects of Levels of Tropical Rice Husk-Derived Biochar in Diet-Based High Rice Straw on In Vitro Methane Production and Rumen Fermentation," *Biomass Conversion and Biorefinery* (2022): 1–7.
10. John Dinsley, *Charcoal Remedies: The Complete Handbook of Medicinal Charcoal and Its Applications* (Coldwater, MI: Gatekeeper Books, 2005).
11. George E. Day, *Swine: A Book for Students and Farmers* (Kenyon Press, 1906), 92.
12. Pennsylvania State College, Agricultural Experiment Station, "Annual Report of the Pennsylvania Agricultural Experiment Station," 1905, 47.
13. Elmer Seth Savage, *Feeding Dairy Cattle* (Holstein-Friesian World, 1917), 47.
14. "Drugs for Specific Purposes in the Ruminant Digestive System," Merck Manual: Veterinary Manual, accessed January 15, 2024, https://www.merckvetmanual.com/pharmacology/systemic-pharmacotherapeutics-of-the-digestive-system/drugs-for-specific-purposes-in-the-ruminant-digestive-system.
15. E. P. Garillo, "Effects of Activated Carbon on Growth, Ruminal Characteristics, Blood Profiles and Feed Digestibility in Sheep," *Asian-Australasian Journal of Animal Sciences* 8 (1995).
16. Miia Lindström et al., "*Clostridium botulinum* in Cattle and Dairy Products," *Critical Reviews in Food Science and Nutrition* 50, no. 4 (2010): 281–304.
17. H. F. Gomez et al., "Adsorption of Botulinum Toxin to Activated Charcoal with a Mouse Bioassay," *Annals of Emergency Medicine* 25, no. 6 (1995): 818–22.
18. M. Kruger et al., "Alternative Methods to Reduce Infectious Burdens in Farm Animal Stables," presented at the Workshop Pharmaceuticals in Soil, Sludge and Slurry of the German Federal Environment Agency, June 18–19, 2013.
19. F. Diez-Gonzalez, "Grain Feeding and the Dissemination of Acid-Resistant *Escherichia coli* from Cattle," *Science* 281, no. 5383 (1998): 1666–68, https://doi.org/10.1126/science.281.5383.1666.
20. K. Naka et al., "Adsorption Effect of Activated Charcoal on Enterohemorrhagic *Escherichia coli*," *The Journal of Veterinary Medical Science* 63, no. 3 (2001): 281–85.

21. J. B. Gurtler, "Inactivation of *E. coli* O157: H7 in Cultivable Soil by Fast and Slow Pyrolysis-Generated Biochar," *Foodborne Pathogens and Disease* 11, no. 3 (2014), 215–23; S. M. Abit et al., "Transport of, and Microspheres in Biochar-Amended Soils with Different Textures," *Journal of Environmental Quality* 43, no. 1 (2014), 371–388; S. M. Abit et al., "Influence of Feedstock and Pyrolysis Temperature of Biochar Amendments on Transport of *Escherichia Coli* in Saturated and Unsaturated Soil," *Environmental Science & Technology* 46, no. 15 (2012): 8097–105.
22. S. Watarai, Tana, and M. Koiwa, "Feeding Activated Charcoal from Bark Containing Wood Vinegar Liquid (Nekka-Rich) Is Effective as Treatment for Cryptosporidiosis in Calves," *Journal of Dairy Science* 91, no. 4 (2008): 1458–63, https://doi.org/10.3168/jds.2007-0406.
23. R.M. Cook and K. A. Wilson, "Removal of Pesticide Residues from Dairy Cattle," *Journal of Dairy Science* 54, no. 5 (1971): 712–18.
24. K. S. N. Prasad and A. Chabra, "Effect of Supplementing Charcoal on the Organochlorine Pesticide Residues in Milk," *Indian Journal of Animal Nutrition* 23, no. 1 (2006): 53–55.
25. P. S. Erickson, N. L. Whitehouse, and M. L. Dunn, "Activated Carbon Supplementation of Dairy Cow Diets: Effects on Apparent Total-Tract Nutrient Digestibility and Taste Preference," *The Professional Animal Scientist* 27, no. 5 (2011): 428–34.
26. R. R. Marquardt and A. A. Frohlich, "A Review of Recent Advances in Understanding Ochratoxicosis," *Journal of Animal Science* 70, no. 12 (1992): 3968–88.
27. Do Thi Thanh Van, "Some Animal and Feed Factors Affecting Feed Intake, Behaviour and Performance of Small Ruminants" (PhD thesis, Swedish University of Agricultural Sciences, 2006); T. N. Barry and T. R. Manley, "The Role of Condensed Tannins in the Nutritional Value of *Lotus Pedunculatus* for Sheep. 2. Quantitative Digestion of Carbohydrates and Proteins," *The British Journal of Nutrition* 51, no. 3 (1984): 493; R. E. Banner et al., "Supplemental Barley and Charcoal Increase Intake of Sagebrush by Lambs," *Journal of Range Management* 53, no. 4 (2000): 415–20.
28. R. Bhatta et al., "Nutrient Content, in Vitroruminal Fermentation Characteristics and Methane Reduction Potential of Tropical Tannin-Containing Leaves," *Journal of the Science of Food and Agriculture* 92, no. 15 (2012): 2929–35, https://doi.org/10.1002/jsfa.5703.
29. Gabriella Cobellis, Massimo Trabalza-Marinucci, and Zhongtang Yu, "Critical Evaluation of Essential Oils as Rumen Modifiers in Ruminant Nutrition: A Review," *Science of the Total Environment* 545 (2016): 556–68.
30. John Dinsley, *Charcoal Remedies: The Complete Handbook of Medicinal Charcoal and Its Applications* (Coldwater, MI: Gatekeeper Books, 2005), 193.
31. John Dinsley, *Charcoal Remedies*, 193.

Chapter 10. Building Soil

1. Johannes Lehmann and Markus Kleber, "The Contentious Nature of Soil Organic Matter," *Nature* 528, no. 7580 (2015): 60–68.
2. Johannes Lehmann et al., "Persistence of Soil Organic Carbon Caused by Functional Complexity," *Nature Geoscience* 13, no. 8 (2020): 529–34.
3. Jianmei Zou et al., "Is the Traditional Alkali Extraction Method Valid in Isolating Chemically Distinct Humic Acid?," *Chemical Engineering Journal Advances* 6 (2021): 100077.

4. Hugh Lovel, *Quantum Agriculture: Biodynamics and Beyond: Growing Plentiful, Vital Food* (Georgia: Quantum Agriculture Publishers, 2014).
5. R. L. Mulvaney, S. A. Khan, and T. R. Ellsworth, "Synthetic Nitrogen Fertilizers Deplete Soil Nitrogen: A Global Dilemma for Sustainable Cereal Production," *Journal of Environmental Quality* 38, no. 6 (2009): 2295–314.
6. Xiangyu Gao et al., "Effects of Alkaline Biochar on Nitrogen Transformation with Fertilizer in Agricultural Soil," *Environmental Research* (2023): 116084.
7. Salar Farhangi-Abriz et al., "A Meta-Analysis to Estimate the Potential of Biochar in Improving Nitrogen Fixation and Plant Biomass of Legumes," *Biomass Conversion and Biorefinery* (2022): 1–11.
8. David Yarrow, "Biochar: Prepping It for Soil," *Acres U.S.A.*, July 2013.
9. Hong Liu et al., "Evaluation of Fungal Dynamics During Sheep Manure Composting Employing Peach Shell Biochar," *Bioresource Technology* (2023): 129559.
10. Nikolas Hagemann et al., "Organic Coating on Biochar Explains Its Nutrient Retention and Stimulation of Soil Fertility," *Nature Communications* 8, no. 1 (2017): 1089; Stephen Joseph et al., "Microstructural and Associated Chemical Changes During the Composting of a High Temperature Biochar: Mechanisms for Nitrate, Phosphate and Other Nutrient Retention and Release," *Science of the Total Environment* 618 (2018): 1210–23.
11. Ehrenfried Pfeiffer, *Chromatography Applied to Quality Testing* (Biodynamic Farming & Gardening Association, 1984).
12. Maria Olga Kokornaczyk et al., "Analysis of Soils by Means of Pfeiffer's Circular Chromatography Test and Comparison to Chemical Analysis Results," *Biological Agriculture & Horticulture* 33, no. 3 (2017): 143–57.
13. W. Brinton, "Assessing Compost & Humus Condition by Circular Chromatography," *Journal of the Woods End Research Lab* 1, no. 1 (2010), https://woodsend.com/publications/.
14. Kenneth Avery and Ann Jackson-Avery, *This Bokashi Alternative Is Rubbish: Bokashi and Pfeiffer Chromatography Simplified* (independently published, 2023).
15. Ben Ford et al., "Using Paper Chromatography for Assessing Soil Health in Southwestern Australia," Centre of Excellence in Natural Resource Management, University of Western Australia, 2019.

Chapter 11. Biochar Recipes

1. Youngsang Cho, *JADAM Organic Farming: The Way to Ultra-Low-Cost Agriculture* (JADAM, 2016).
2. Sandor Katz, *The Art of Fermentation* (White River Junction, VT: Chelsea Green, 2012).
3. Mujahid Rasool, Adnan Akhter, and Muhammad Saleem Haider, "Molecular and Biochemical Insight into Biochar and *Bacillus Subtilis* Induced Defense in Tomatoes against *Alternaria solani*," *Scientia Horticulturae* 285 (2021): 110203.
4. Steve Solomon, *Gardening When It Counts: Growing Food in Hard Times* (Gabriola, BC: New Society Publisher, 2006).
5. Giuseppina Iacomino et al., "The Suppressive Effects of Biochar on Above- and Belowground Plant Pathogens and Pests: A Review," *Plants* 11, no. 22 (2022): 3144.
6. Yanfang Wang et al., "Effects of Biochar on the Growth of Apple Seedlings, Soil Enzyme Activities and Fungal Communities in Replant Disease Soil," *Scientia*

Horticulturae 256 (2019): 108641; Amaya Atucha and Greg Litus, "Effect of Biochar Amendments on Peach Replant Disease," *HortScience* 50, no. 6 (2015): 863–68.
7. Michael Phillips, *The Holistic Orchard: Tree Fruits and Berries the Biological Way* (White River Junction, VT: Chelsea Green, 2012).
8. "Progress Report on Bexley's Charter Oak—5 Years On," Carbon Gold, https://www.carbongold.com/progress-report-on-bexleys-charter-oak/.
9. David Yarrow, "Down the Wormhole: Customizing Biological Methods for Large-Scale Farming," *Acres U.S.A.* 46, no. 8 (August 2016), https://www.ecofarmingdaily.com/wp-content/uploads/August16_Down-the-Wormhole.pdf.

Chapter 12. For the Forest

1. Julie E. Korb, Nancy C. Johnson, and W. W. Covington, "Slash Pile Burning Effects on Soil Biotic and Chemical Properties and Plant Establishment: Recommendations for Amelioration," *Restoration Ecology* 12, no. 1 (2004): 52–62.
2. James E. Amonette et al., *Biomass to Biochar: Maximizing the Carbon Value*, report by Center for Sustaining Agriculture and Natural Resources, Washington State University, Pullman, WA, 2021, https://csanr.wsu.edu/biomass2biochar/.
3. Matthew W. Jones et al., "Global Fire Emissions Buffered by the Production of Pyrogenic Carbon," *Nature Geoscience* 12, no. 9 (2019): 742–47.
4. Moritz Reisser et al., "Pyrogenic Carbon in Soils: A Literature-Based Inventory and a Global Estimation of Its Content in Soil Organic Carbon and Stocks," *Frontiers in Earth Science* 4 (2016): 80.
5. Thomas H. DeLuca and Gregory H. Aplet, "Charcoal and Carbon Storage in Forest Soils of the Rocky Mountain West," *Frontiers in Ecology and the Environment* 6, no. 1 (2008): 18–24.
6. Cristina Santín et al., "Pyrogenic Organic Matter Production from Wildfires: A Missing Sink in The Global Carbon Cycle," *Global Change Biology* 21, no. 4 (2015): 1621–33.
7. S. A. Hart and N. J. Luckai, "Charcoal Carbon Pool in North American Boreal Forests," *Ecosphere* 5, no. 8 (2014): 1–14.
8. Beatriz Omil, Verónica Piñeiro, and Agustín Merino, "Soil and Tree Responses to the Application of Wood Ash Containing Charcoal in Two Soils with Contrasting Properties," *Forest Ecology and Management* 295 (2013): 199–212.
9. Michael J. Gundale and Thomas H. DeLuca, "Temperature and Source Material Influence Ecological Attributes of Ponderosa Pine and Douglas-Fir Charcoal," *Forest Ecology and Management* 231, no. 1–3 (2006): 86–93.
10. L. M. Berglund, T. H. DeLuca, and O. Zackrisson, "Activated Carbon Amendments to Soil Alters Nitrification Rates in Scots Pine Forests," *Soil Biology and Biochemistry* 36, no. 12 (2004): 2067–73.
11. Johannes Lehmann et al., "Biochar Effects on Soil Biota—A Review," *Soil Biology and Biochemistry* 43, no. 9 (2011): 1812–36.
12. K. Makoto et al., "Buried Charcoal Layer and Ectomycorrhizae Cooperatively Promote the Growth of *Larix gmelinii* Seedlings," *Plant and Soil* 327 (2010): 143–52.
13. Caroline A. Masiello et al., "Biochar Effects on Soil Hydrology," in *Biochar for Environmental Management*, eds. Johannes Lehmann and Stephen Joseph (Routledge, 2015), 543–62.

14. Earl Haven Tryon, "Effect of Charcoal on Certain Physical, Chemical, and Biological Properties of Forest Soils," *Ecological Monographs* 18, no. 1 (1948): 81–115.
15. Sean C. Thomas and Nigel Gale, "Biochar and Forest Restoration: A Review and Meta-Analysis of Tree Growth Responses," *New Forests* 46, no. 5–6 (2015): 931–46.
16. Rachel J. Brimmer, "Charcoal Quantity and Adsorptive Activity in Ponderosa Pine Ecosystems of Western Montana," MS thesis, University of Montana, 2006.
17. Thomas H. DeLuca and Gregory H. Aplet, "Charcoal and Carbon Storage in Forest Soils of the Rocky Mountain West," *Frontiers in Ecology and the Environment* 6, no. 1 (2008): 18–24.
18. Verena Jauss et al., "Pyrogenic Carbon Controls Across a Soil Catena in the Pacific Northwest," *Catena* 124 (2015): 53–59.
19. Heike Knicker et al., "Characteristic Alterations of Quantity and Quality of Soil Organic Matter Caused by Forest Fires in Continental Mediterranean Ecosystems: A Solid State 13C NMR Study," *European Journal of Soil Science* 57, no. 4 (2006): 558–69.
20. C. Santín et al., "Consumption of Residual Pyrogenic Carbon by Wildfire," *International Journal of Wildland Fire* 22, no. 8 (2013): 1072–77.
21. Eric Oliveira Carvalho, Leda N. Kobziar, and Francis E. Putz, "Fire Ignition Patterns Affect Production of Charcoal in Southern Forests," *International Journal of Wildland Fire* 20, no. 3 (2011): 474–77.
22. Jonathan W. Long, Frank K. Lake, and Ron W. Goode, "The Importance of Indigenous Cultural Burning in Forested Regions of the Pacific West, USA," *Forest Ecology and Management* 500 (2021): 119597.
23. Maureen Puettmann, Kelpie Wilson, and Elaine Oneil, "Life Cycle Assessment of Biochar from Postharvest Forest Residues," draft report for Consortium for Research on Renewable Industrial Materials, Corvallis, OR, 2017.
24. Jiøí Woitsch, "Charcoal Makers in Bohemia: From Privileged Craftsmen to Strange Forest Dwellers," in *Woodland Cultures in Time and Space: Tales from the Past, Messages for the Future*, ed. Eirini Saratsi (Embryo Publications, 2009).
25. Kelpie J. Wilson, "A Carbon Conservation Corps to Restore Forests with Biochar Using Flame-Cap Kilns," at 2021 ASABE Annual International Virtual Meeting, American Society of Agricultural and Biological Engineers, 2021.
26. Charles E. Heller, "The US Army, the Civilian Conservation Corps, and Leadership for World War II, 1933–1942," *Armed Forces & Society* 36, no. 3 (2010): 439–53.
27. Tomi Hazel Vaarde, *Social Forestry: Tending the Land as People of Place* (Santa Fe, NM: Bluebird Press, 2023).
28. "Innovations in Biochar," USDA Natural Resources Conservation Service, October 22, 2019, https://www.nrcs.usda.gov/conservation-basics/conservation-by-state/oregon/news/innovations-in-biochar.
29. Aldo Leopold, *A Sand County Almanac* (Oxford University Press, 1949).

Chapter 13. Counting Carbon

1. Albert Bates and Kathleen Draper, *Burn: Using Fire to Cool the Earth* (White River Junction, VT: Chelsea Green, 2018).
2. Gerard Cornelissen et al., "Flame Curtain Kilns Produce Biochar from Dry Biomass with Minimal Methane Emissions," *Science of The Total Environment* (2023): 166547.

3. K. J. Wilson, W. Bekker, and S.I. Feher, "Producing, Characterizing and Quantifying Biochar in the Woods Using Portable Flame Cap Kilns," *Journal of Visualized Experiments* 203 (2024): e65543, https://doi.org/10.3791/65543.
4. Lin Wang, Dingjiang Chen, and Lizhong Zhu, "Biochar Carbon Sequestration Potential Rectification in Soils: Synthesis Effects of Biochar on Soil CO_2, CH_4 And N_2O Emissions," *Science of The Total Environment* 904 (2023): 167047.

Chapter 14. Return to the Garden of Eatin'

5. Jules Bernstein, "Grow and Eat Your Own Vaccines?" UC Riverside, September 16, 2021. https://www.universityofcalifornia.edu/news/grow-and-eat-your-own-vaccines.
6. "Please Pass the Salad—Edible Vaccines Produced in Lettuce to Protect Against COVID-19," University of Ottawa, May 9 2022, https://www.uottawa.ca/faculty-science/please-pass-salad-edible-vaccines-produced-lettuce-protect-against-covid-19.
7. *The State of the World's Land and Water Resources for Food and Agriculture—Systems at Breaking Point, Synthesis Report 2021* (Rome: FAO, 2021), https://doi.org/10.4060/cb7654en.
8. H. Eswaram, R. Lal, and P. F. Reich, "Land Degradation: An Overview," presented at the Responses to Land Degradation, proceedings of the 2nd International Conference on Land Degradation and Desertification, Khon Kaen, Thailand, 2001.
9. Rattan Lal, "Depletion and Restoration of Carbon in the Pedosphere (< Special Issue> International Symposium: Challenges to Soil Degradation Towards Sustaining Life and Environment, Tokyo Metropolitan University Symposium Series No. 2, 2009)," *Pedologist* 53, no. 3 (2010): 19-32.
10. Xiang-Hong Liu, Feng-Peng Han, and Xing-Chang Zhang, "Effect of Biochar on Soil Aggregates in the Loess Plateau: Results from Incubation Experiments," *International Journal of Agriculture & Biology* 14, no. 6 (2012): 975–79.
11. J. H. Yuan, & R. K. Xu, "The Amelioration Effects of Low Temperature Biochar Generated from Nine Crop Residues on an Acidic Ultisol," *Soil Use and Management* 27, no. 1 (March 2011), https://doi.org/10.1111/j.1475-2743.2010.00317.x.
12. S. Supriyadi et al., "Effect of Biochar on P Uptake from Two Acid Soils," proceedings of the 16th Australian Agronomy Conference, 2012.
13. J. M. Novak et al., "Impact of Biochar Amendment on Fertility of a Southeastern Coastal Plain Soil," *Soil Science* 174, no. 2 (2009): 105–112, https://doi.org/10.1097/SS.0b013e3181981d9a.
14. S. D. Joseph et al., "An Investigation into the Reactions of Biochar in Soil," *Australian Journal of Soil Research* 48, no.7 (2010): 501, https://doi.org/10.1071/SR10009.
15. X.-H. Liu and X.-C. Zhang, "Effect of Biochar on ph of Alkaline Soils in the Loess Plateau: Results from Incubation Experiments," *International Journal Of Agriculture & Biology* 14, no. 5 (2012): 745–50.
16. P. Barak, "Saline and Sodic Soils in Wisconsin: The Sad Story of an Urban Infiltration Pond," presented at the Waukesha County Storm Water Workshop, 2012.
17. M. S. Lashari et al., "Effects of Amendment of Biochar-Manure Compost in Conjunction with Pyroligneous Solution on Soil Quality and Wheat Yield of a Salt-Stressed Cropland from Central China Great Plain," *Field Crops Research* 144 (2013): 113–18, https://doi.org/10.1016/j.fcr.2012.11.015.

18. Yanfei Yuan et al., "Biochar as a Sustainable Tool for Improving the Health of Salt-Affected Soil," *Soil & Environmental Health* (2023): 100033.
19. L. Beesley, "A Review of Biochars' Potential Role in the Remediation, Revegetation and Restoration of Contaminated Soils," *Environmental Pollution* 159, no. 12 (2011): 3269–82, https://doi.org/10.1016/j.envpol.2011.07.023.
20. L. Beesley et al., "Biochar Addition to an Arsenic Contaminated Soil Increases Arsenic Concentrations in the Pore Water but Reduces Uptake to Tomato Plants (*Solanum lycopersicum L.*)," *Science of the Total Environment* (2013): 454–55, 598–603, https://doi.org/10.1016/j.scitotenv.2013.02.047.
21. M. Ahmad et al., "Effects of Soil Dilution and Amendments (Mussel Shell, Cow Bone, and Biochar) on Pb Availability and Phytotoxicity in Military Shooting Range Soil," *Ecotoxicology and Environmental Safety* 79 (May 2012): 225–31, www.sciencedirect.com/S0147651312000073.
22. J. Beniston and R. Lal, "Improving Soil Quality for Urban Agriculture in the North Central US," in *Carbon Sequestration in Urban Ecosystems* (Springer Netherlands, 2012), 279–313.
23. Y. Wang et al., "Reducing the Bioavailability of Pcbs in Soil to Plant by Biochars Assessed with Triolein-Embedded Cellulose Acetate Membrane Technique," *Environmental Pollution* 174, no. C (2013), 250–56, https://doi.org/10.1016/j.envpol.2012.12.004.
24. X. Zhang et al., "Using Biochar for Remediation of Soils Contaminated with Heavy Metals and Organic Pollutants," *Environmental Science and Pollution Research* 20, no. 12 (2013): 8472–83, link.springer.com/s11356-013-1659-0.
25. X. B. Yang et al., "Influence of Biochars on Plant Uptake and Dissipation of Two Pesticides in an Agricultural Soil," *Journal of Agricultural and Food Chemistry* 58, no. 13 (2010): 7915–21, https://doi.org/10.1021/jf1011352.
26. K. M. Bushnaf et al., "Effect of Biochar on the Fate of Volatile Petroleum Hydrocarbons in an Aerobic Sandy Soil," *Journal of Contaminant Hydrology* 126, no. 3–4 (2011): 208–15, https://doi.org/10.1016/j.jconhyd.2011.08.008.
27. G. Fellet et al., "Application of Biochar on Mine Tailings: Effects and Perspectives for Land Reclamation," *Chemosphere* 83, no. 9 (May 2011): 1262–67, https://doi.org/10.1016/j.chemosphere.2011.03.053.
28. Daniel Hillel, *Out of the Earth: Civilization and the Life of the Soil* (University of California Press, 1992).
29. Gail Tverberg, "Models Hide the Shortcomings of Wind and Solar," *Our Finite World* (blog), June 2, 2023, https://ourfiniteworld.com/2023/06/02/models-hide-the-shortcomings-of-wind-and-solar/.
30. Nathan J. Hagens, "Economics for the Future—Beyond the Superorganism," *Ecological Economics* 169 (2020): 106520.
31. "Japan's Sustainable Society in the Edo Period (1603-1867)," Japan for Sustainability newsletter, no. 7 (March 2003), https://www.japanfs.org/sp/en/news/archives/news_id027757.html.
32. Warren Johnson, *Muddling Toward Frugality* (Easton Studio Press LLC, 2010).
33. Masanobu Fukuoka, *The Natural Way of Farming: The Theory and Practice of Green Philosophy* (Tokyo and New York: Japan Publications, Inc., 1985), 263.

Index

Page numbers followed by *f* or *t* indicate figures and tables, respectively.

Index

Index

Index

Index

Index

Index

About the Author

Daniel Sharp Photography

KELPIE WILSON is a mechanical engineer and analyst with decades of experience in renewable energy, sustainable forestry, and resource conservation. From 2008 to 2012 Kelpie worked for the International Biochar Initiative as a project developer and managed the multistakeholder process to create the first set of standards and testing guidelines for biochar. She was also a founding board member of the US Biochar Initiative. Since 2012, she has been a consultant helping biochar companies, NGOs, and government agencies understand biochar markets and production technologies. She has authored and coauthored many publications on different aspects of the biochar industry, including both biochar production systems and biochar applications.

Kelpie researches and teaches techniques for making and using biochar in compost and manure management, and presents many biochar workshops each year to farmers, foresters, resource conservation districts, fire safe councils, and landowners. She also develops systems for making biochar from waste biomass. She pioneered development of the flame-cap kiln biochar technology for use by forestry contractors, arborists, and farmers, and she manufactures and sells the Ring of Fire Biochar Kiln® through Wilson Biochar, LLC. She manages the Biochar in the Woods network, sponsored by the US Biochar Initiative, bringing together forestry contractors, agencies, and small woodland owners to share knowledge and approaches for converting forest slash into biochar.